Lecture Notes in Mathematics

Volume 2359

Editors-in-Chief

Jean-Michel Morel, City University of Hong Kong, Kowloon Tong, China
Bernard Teissier, IMJ-PRG, Paris, France

Series Editors

Karin Baur, University of Leeds, Leeds, UK
Michel Brion, UGA, Grenoble, France
Rupert Frank, LMU, Munich, Germany
Annette Huber, Albert Ludwig University, Freiburg, Germany
Davar Khoshnevisan, The University of Utah, Salt Lake City, UT, USA
Ioannis Kontoyiannis, University of Cambridge, Cambridge, UK
Angela Kunoth, University of Cologne, Cologne, Germany
Ariane Mézard, IMJ-PRG, Paris, France
Mark Podolskij, University of Luxembourg, Esch-sur-Alzette, Luxembourg
Mark Policott, Mathematics Institute, University of Warwick, Coventry, UK
László Székelyhidi ⓘ, MPI for Mathematics in the Sciences, Leipzig, Germany
Gabriele Vezzosi, UniFI, Florence, Italy
Anna Wienhard, MPI for Mathematics in the Sciences, Leipzig, Germany

This series reports on new developments in all areas of mathematics and their applications - quickly, informally and at a high level. Mathematical texts analysing new developments in modelling and numerical simulation are welcome. The type of material considered for publication includes:

1. Research monographs
2. Lectures on a new field or presentations of a new angle in a classical field
3. Summer schools and intensive courses on topics of current research.

Texts which are out of print but still in demand may also be considered if they fall within these categories. The timeliness of a manuscript is sometimes more important than its form, which may be preliminary or tentative. Please visit the LNM Editorial Policy (https://drive.google.com/file/d/1MOg4TbwOSokRnFJ3ZR3ciEeKs9hOnNX_/view?usp=sharing)

Titles from this series are indexed by Scopus, Web of Science, Mathematical Reviews, and zbMATH.

Yuri Kordyukov • Fedor Sukochev • Dmitriy Zanin

Principal Symbol Calculus on Contact Manifolds

Yuri Kordyukov
Institute of Mathematics
Ufa Federal Research Centre of Russian
Academy of Sciences
Ufa, Russia

Fedor Sukochev
School of Mathematics & Statistics
UNSW Sydney
Sydney, NSW, Australia

Dmitriy Zanin
School of Mathematics & Statistics
UNSW Sydney
Sydney, NSW, Australia

ISSN 0075-8434 ISSN 1617-9692 (electronic)
Lecture Notes in Mathematics
ISBN 978-3-031-69925-2 ISBN 978-3-031-69926-9 (eBook)
https://doi.org/10.1007/978-3-031-69926-9

Mathematics Subject Classification: 46-xx, 47-xx, 53-xx

© The Editor(s) (if applicable) and The Author(s), under exclusive license to Springer Nature Switzerland AG 2024

This work is subject to copyright. All rights are solely and exclusively licensed by the Publisher, whether the whole or part of the material is concerned, specifically the rights of translation, reprinting, reuse of illustrations, recitation, broadcasting, reproduction on microfilms or in any other physical way, and transmission or information storage and retrieval, electronic adaptation, computer software, or by similar or dissimilar methodology now known or hereafter developed.
The use of general descriptive names, registered names, trademarks, service marks, etc. in this publication does not imply, even in the absence of a specific statement, that such names are exempt from the relevant protective laws and regulations and therefore free for general use.
The publisher, the authors and the editors are safe to assume that the advice and information in this book are believed to be true and accurate at the date of publication. Neither the publisher nor the authors or the editors give a warranty, expressed or implied, with respect to the material contained herein or for any errors or omissions that may have been made. The publisher remains neutral with regard to jurisdictional claims in published maps and institutional affiliations.

This Springer imprint is published by the registered company Springer Nature Switzerland AG
The registered company address is: Gewerbestrasse 11, 6330 Cham, Switzerland

If disposing of this product, please recycle the paper.

To Nigel Higson, who saw the path.

Foreword

It is a pleasure to contribute a few words to this monograph by Yuri Kordyukov, Fedor Sukochev and Dima Zanin on the operator theoretic, or C^*-algebraic, approach to the Heisenberg pseudodifferential calculus on contact manifolds. This topic has attracted increasing attention over the past 20 years, and at the present time researchers with a broad range of backgrounds, techniques and perspectives are at work on it, or on very closely related problems. For that reason, this monograph, which addresses foundational questions about the principal symbols of pseudodifferential operators, will be welcomed by many.

A noteworthy early achievement, bringing together contact manifolds, pseudodifferential operators and C^*-algebras, was Erik van Erp's formulation and proof of the index theorem for Heisenberg-elliptic operators on a contact manifold (*Ann. of Math.*, 2010).

Van Erp made the crucial observation that, for the purposes of index theory, the symbol of a Heisenberg-elliptic operator should be treated as an element in the K-theory of a noncommutative C^*-algebra, namely the C^*-algebra of continuous sections of the same bundle of nilpotent groups that is described in Sect. 1.3 of the present monograph. Earlier attempts to develop index theory on contact manifolds struggled with the issue of the symbol class, but with van Erp's discovery in hand, the various component parts of Atiyah-Singer index theory could be transferred in a very conceptual way to the Heisenberg context, leading quite quickly to a complete theory on contact manifolds.

Van Erp's discoveries gave prominence to the idea that C^*-algebras have a natural role to play as carriers of the symbol, and this in turn led to a range of natural questions in C^*-algebra theory, many of which are investigated and indeed answered in the present work, which is organized around the detailed construction of a principal symbol homomorphism from the C^*-algebra generated by order-zero operators in the Heisenberg calculus to van Erp's C^*-algebra of principal symbols. Fundamental to everything is the C^*-algebraic short exact sequence described in Theorem 1.3.18 in this monograph. Kordyukov, Sukochev and Zanin present a new and thoroughly detailed account of this result, built from the ground up using self-adjoint operator theory.

It is natural to ask if one can go beyond index theory and topology, and investigate geometric questions on contact or other sub-Riemmannian manifolds from the perspective of operator theory. An obvious starting point, which is arguably the starting point for all of geometric analysis and all of noncommutative geometry, is Weyl's formula, relating volume to the growth of eigenvalues of the Laplacian.

The connection between Weyl's formula, algebras of pseudodifferential operators and traces was made a long time ago by Victor Guillemin (*Adv. in Math.*, 1985). Work of Alain Connes from around the same time (*Comm. Math. Phys.*, 1988) emphasized the role of the Dixmier trace and offered an interesting change in perspective, involving the development of notions of volume and integral in contexts far beyond the conventional. Connes' so-called trace theorem connects his ideas about volume and integral with the usual ones in the conventional context. Sukochev and Zanin, in collaboration with others, have made numerous contributions to the precise formulation and proof of Connes' trace theorem, and to a number of challenging refinements, and the present monograph studies the Heisenberg case in detail; see Theorem 1.4.6. Even though the Heisenberg case is arguably not far from the conventional case, relative to the broad range of possibilities that Connes' theory affords, there are still surprises in the final formula.

Let me close by mentioning two prospects for the future. First, there is much more within sub-Riemannian geometry that remains to be understood from the operator-theoretic perspective. Connes' concept of spectral triple offers one starting point for the exploration of geometric spaces beyond the context of Riemannian geometry (and indeed his ideas about integral and volume were developed within the framework of spectral triples). But the seemingly small amount on noncommutativity that is introduced through the osculating nilpotent groups of a contact manifold, or other sub-Riemannian space, seems to fit only awkwardly into the spectral triple framework, and something a bit different may eventually be required.

Second, in recent breakthrough work of Iakovos Androulidakis, Omar Mohsen and Bob Yuncken, on pseudodifferential operators associated to manifold structures that are more general and more singular than contact structures, C^*-algebras again play a decisive role, but now in a different way and in a completely different area: the study of maximal hypoellipticity for linear partial differential operators. The C^*-algebras and C^*-algebra extensions that arise here are very new and very little explored, so far. There is much more work to be done!

July 6, 2024   Nigel Higson

Preface

The idea of the principal symbol is fundamental in the theory of elliptic operators, and by extension in the study of partial differential equations. A differential operator P of order m on $\mathbb{R}^d$ can generically be written in the form

$$P = \sum_{|\alpha| \leq m} a_\alpha(x) \partial_x^\alpha,$$

where the sum is over multi-indices $\alpha = (\alpha_1, \cdots, \alpha_d)$ with $|\alpha| = \sum_{j=1}^d \alpha_j \leq m$, the coefficients a_α are smooth functions of $x \in \mathbb{R}^d$ and ∂_x^α denotes the differential operator $\partial_{x_1}^{\alpha_1} \cdots \partial_{x_d}^{\alpha_d}$.

Given a partial differential operator P in the space $\mathbb{R}^d$, we want to analyse the equation $Pu = f$, where f is some known function and u is an unknown distribution. In general, it can be very difficult to determine the existence of solutions u; however, in the case when P has constant coefficients (or equivalently, P is invariant under the group of translations of $\mathbb{R}^d$) $Pu = f$ can be transformed into an algebraic equation using the Fourier transform. When P is not translation invariant, more sophisticated methods are needed. One of the most powerful techniques are the a priori estimates, which are estimates for the size and regularity of u. In general, no such estimates are possible: for example, if $P = 0$ and $f = 0$, then the equation $Pu = f$ says nothing about u. We can make progress in the case that P is elliptic.

The *principal symbol* of P is the function $\sigma(P)$ defined by

$$\sigma(P)(x, \xi) = \sum_{|\alpha| = m} i^m a_\alpha(x) \xi^\alpha,$$

where $\xi \in \mathbb{R}^d$ and $\xi^\alpha = \xi_1^{\alpha_1} \cdots \xi_d^{\alpha_d}$. The principal symbol contains important information about the nature of the solutions to the equation $Pu = f$ for an unknown function u. For example, if $\sigma(P)(x, \xi)$ is invertible away from $\xi = 0$, then P is said to be elliptic at x and if f is smooth at x then so is u. The basic idea is that while it may not be possible to invert P and write $u = P^{-1}f$, if $\sigma(P)(x, \xi)$

is invertible away from $\xi = 0$, then the function

$$\widetilde{u}(x) = (2\pi)^{-d} \int_{\mathbb{R}^d} e^{i(x,\xi)} \sigma(P)(x,\xi)^{-1} \psi(\xi) \widehat{f}(\xi)\, d\xi$$

forms a good approximation for u in the sense that $u - \widetilde{u}$ is expected to be smoother than u. Here, ψ is a smooth function which vanishes near 0 and is equal to 1 at infinity, which is needed to eliminate the singularity in $\sigma(P)(x,\xi)^{-1}$ at $\xi = 0$.

These considerations lead to the theory of pseudodifferential operators on $\mathbb{R}^d$. The algebra $\Psi(\mathbb{R}^d)$ of classical pseudodifferential operators is a filtered algebra $\{\Psi^k(\mathbb{R}^d)\}_{k \in \mathbb{Z}}$ of linear endomorphisms of the Schwartz space $\mathcal{S}(\mathbb{R}^d)$. The idea is that given suitable regularity properties on the coefficients a_α, the differential operator P belongs to $\Psi^m(\mathbb{R}^d)$, and there is an approximate inverse of P belonging to $\Psi^{-m}(\mathbb{R}^d)$. The operators belonging to $\Psi^k(\mathbb{R}^d)$ have a principal symbol function of $(x,\xi) \in \mathbb{R}^d \times (\mathbb{R}^d \setminus \{0\})$ which is a homogeneous function of order k in ξ. In particular, the elements of $\Psi^0(\mathbb{R}^d)$ have a principal symbol which can be identified with a smooth function on $\mathbb{R}^d \times S^{d-1}$.

The elements of $\Psi^0(\mathbb{R}^d)$ define bounded operators on the Hilbert space $L_2(\mathbb{R}^d)$ of square-integrable functions on $\mathbb{R}^d$, and the principal symbol mapping is an algebra homomorphism from $\Psi^0(\mathbb{R}^d)$ to $C^\infty(\mathbb{R}^d \times S^{d-1})$. An understanding of Ψ^0 *from the point of view of operator theory* was obtained by Beals in [6], and the principal symbol mapping was defined from the C^*-algebraic perspective by Cordes [15].

What we have attempted in [43, 47, 59] is a fully synthetic theory of the C^*-closure of the algebra of classical pseudodifferential operators of order 0. Here, synthetic means that we avoid the explicit use of the Fourier transform and instead we define Ψ^0 as an algebra generated by a specific family of elementary operators. There is also no explicit reference to partial differential operators: all operators from the beginning are bounded on L_2. The starting point of [47, 59] was noting that the most basic pseudodifferential operators of order 0 are the *pointwise multipliers* and the *dilation-invariant Fourier multipliers*, and that together the C^*-algebra generated by these operators is precisely the C^*-closure of the algebra of pseudodifferential operators of order 0. The motivating application for [47, 59] was to find a new proof of Connes' trace theorem, which is a statement relating the asymptotic behaviour of the eigenvalues of negative order pseudodifferential operators with their principal symbol.

The notion of the principal symbol of a differential operator extends to differential operators on manifolds in a natural way: a differential operator on a manifold is one that coincides with a differential operator in every coordinate chart, and the principal symbol transforms covariantly so as to define a smooth function on the cotangent bundle. As with operators on $\mathbb{R}^d$, the algebra of differential operators on a manifold can be extended to a filtered algebra of pseudodifferential operators and these operators have a principal symbol. The theory of [47, 59] was extended to characterise the order zero operators and their principal symbols in [43].

There are many situations where the traditional definition of the principal symbol is not appropriate because the lower order terms in the definition of the operator P contribute in a nontrivial way. An important class of examples occurs in the context of contact manifolds. A contact manifold is a manifold X whose tangent bundle TX is equipped with a co-rank 1 sub-bundle H which is completely non-integrable. Complete non-integrability means approximately that if X and Y are sections of H, then their commutator $[X, Y]$ is not a section of H. The differential calculus associated to contact manifold is defined so that differentiations in the directions of H count as order 1 operators, but differentiation in other directions defines an operator of order 2.

In the present paper we extend [43, 47, 59] further to the cutting edge of pseudodifferential operator theory in order to better understand the order zero operators in the Heisenberg calculus on contact manifolds from the point of view of operator theory. Just as a smooth manifold is locally modelled on $\mathbb{R}^d$, a contact manifold is correctly understood as being locally modelled on a Heisenberg group. Therefore, our starting point will be to define an algebra of operators acting on functions on Heisenberg groups.

In Chap. 1, we give a more detailed introduction to the ideas of this paper, including a full definition of contact structure. This includes an overview of the preexisting theories of pseudodifferential calculus on contact manifold and the relation to the work here.

In Chap. 2, we define the principal symbol for operators on the Heisenberg group defined by our calculus. This involves a careful study of elliptic operators in the natural calculus of differential operators on Heisenberg groups. This includes the proof of a priori estimates for elliptic operators. An important development here is the proof of a local Connes trace theorem in the setting of the Heisenberg group.

In Chap. 3, we show how the principal symbol defined in Chap. 2 varies under contact diffeomorphism. This is the most technical part of the paper.

In Chap. 4, we use the results of Chap. 3 to define the principal symbol of pseudodifferential operators on contact manifolds. The signature application of this theory is a proof of a general version of Connes' trace formula for operators on contact manifolds. This version of Connes' trace theorem uses the new principal symbol mapping developed here and is correctly stated in terms of the *Liouville trace on the group von Neumann algebra bundle*, a noncommutative analogy of the Liouville measure on the cotangent bundle defined in Sect. 4.3.

Ufa, Russia Yuri Kordyukov
Sydney, Australia Fedor Sukochev
Sydney, Australia Dmitriy Zanin

Contents

1 **Introduction** 1
   1.1 C^*-Algebraic Approach to the Principal Symbol Mapping
       on Smooth Manifolds 1
   1.2 Contact Manifolds 5
   1.3 Principal Symbol Mapping on Contact Manifolds 6
   1.4 Connes Trace Theorem in the Setting of Contact Manifolds 14

2 **Principal Symbol on the Heisenberg Group** 19
   2.1 Preliminaries on Heisenberg Group 19
       2.1.1 Schatten Class, Ideals and Traces 19
       2.1.2 Non-commutative L_p Spaces and $L_{p,\infty}$ Spaces 20
       2.1.3 Group von Neumann Algebra 21
       2.1.4 Schrödinger Representation of $\mathbb{H}^d$ 22
       2.1.5 Homogeneous Group von Neumann Algebra 24
       2.1.6 Automorphisms of the Heisenberg Groups 26
       2.1.7 Sobolev Spaces 28
   2.2 Existence of the Principal Symbol Map............................... 29
   2.3 Connes Trace Formula for the Heisenberg Sub-Laplacian 34
   2.4 Elliptic Estimate... 41
   2.5 Self-Adjointness of Second Order Differential Operators 49
   2.6 Local Connes Trace Formula 57

3 **Equivariance of the Principal Symbol Under Heisenberg Diffeomorphisms** 65
   3.1 The Action on the Codomain of the Principal Symbol Map 65
       3.1.1 Proof of Theorem 3.1.1 66
       3.1.2 Square Root Lemmas 71
       3.1.3 Continuity Theorem 76
       3.1.4 Proof of Theorem 3.1.2 81
   3.2 Main Computational Theorem 81
       3.2.1 Continuity Theorem 82
       3.2.2 Compactness Theorems 87

		3.2.3	Approximation Theorem	91
		3.2.4	Proof of Theorem 3.2.2	96
	3.3	Invariance of Principal Symbol Map		100
		3.3.1	Conjugation of the Differential Calculus by the Heisenberg Diffeomorphisms	100
		3.3.2	Equivariance Result on the Heisenberg Group	102
		3.3.3	Equivariant Behaviour of the Principal Symbol Under Global Diffeomorphisms	106
		3.3.4	Equivariant Behaviour of the Principal Symbol Under Local Diffeomorphisms	113
4	**Principal Symbol on Contact Manifolds**			**117**
	4.1	Principal Symbol on Compact Contact Manifolds		117
		4.1.1	Globalisation Theorem	117
		4.1.2	Construction of the Principal Symbol Mapping	118
		4.1.3	Proof of Theorem 4.1.8	120
	4.2	Sub-Laplacian on a Compact Contact Manifold		126
		4.2.1	Sobolev Spaces on Compact Contact Manifolds	126
		4.2.2	Definition of Sub-Laplacian	126
		4.2.3	Proof of Theorem 4.2.3	127
		4.2.4	Ellipticity of the Sub-Laplacian	130
		4.2.5	Self-Adjointness of the Sub-Laplacian	133
	4.3	Liouville Trace on the Group von Neumann Algebra Bundle		134
	4.4	Proof of Theorem 1.4.6		136
	4.5	Spectrally Correct Sub-Riemannian Volume		147

References 151

Chapter 1
Introduction

1.1 C^*-Algebraic Approach to the Principal Symbol Mapping on Smooth Manifolds

A central notion of the theory of pseudodifferential operators (PSDOs) is that of a principal symbol, which is roughly a homomorphism from the algebra of pseudo-differential operators into an algebra of functions, [41, Lemma 5.1], [39, Theorem 5.5], [63, pp. 54–55]. Usually, it is defined in a manner inhospitable for operator theorists. However, in [59], a new approach (in the setting of PSDOs on $\mathbb{R}^d$) to a principal symbol mapping on a certain C^*-subalgebra Π in $B(L_2(\mathbb{R}^d))$ is proposed; this mapping turns out to be a $*$-homomorphism from Π into a commutative C^*-algebra. The C^*-algebra Π contains all classical compactly based pseudodifferential operators on $\mathbb{R}^d$. This provides a C^*-algebraic approach to the theory which is arguably more "user friendly". The latter theory was extended to the setting of smooth compact manifolds in [43].

Whereas the approach in [59] is more elementary than the classical approach, the C^*-algebra Π introduced in [59] (see also [47]) is much wider than the class of a classical compactly based pseudo-differential operators of order 0 on $\mathbb{R}^d$. Here comes the key point: the C^*-algebra Π is the closure (in the uniform norm) of the $*$-algebra of all compactly supported *classical* pseudodifferential operators of order 0 (on $\mathbb{R}^d$). However, we use an elementary definition of Π which does not involve pseudodifferential operators. The idea to consider this closure may be discerned yet in [3] (see Proposition 5.2 on p.512). For the recent development of this idea we refer to [47, 59].

Let us explain the definition of Π and that of the principal symbol mapping in the setting of Euclidean space $\mathbb{R}^d$. Let $D_k = \frac{\partial}{\iota \partial t_k}$ be the $k-$th partial derivative operator on $\mathbb{R}^d$ (these are unbounded self-adjoint operators on $L_2(\mathbb{R}^d)$ (here, $\iota = \sqrt{-1}$). Denote $\nabla = (D_1, \cdots, D_d)$ and $\Delta = \sum_{k=1}^d D_k^2$. Let the $d-$dimensional vector $\frac{\nabla}{\Delta^{\frac12}}$ be defined by the functional calculus (its spectrum is the unit sphere $\mathbb{S}^{d-1}$ and its spectral measure is absolutely continuous with respect to the Lebesgue measure on the sphere). Let M_f be the multiplication operator by a function f.

Definition 1.1.1 Let $\pi_1 : L_\infty(\mathbb{R}^d) \to B(L_2(\mathbb{R}^d))$, $\pi_2 : L_\infty(\mathbb{S}^{d-1}) \to B(L_2(\mathbb{R}^d))$ be defined by setting

$$\pi_1(f) = M_f, \quad \pi_2(g) = g(\frac{\nabla}{\sqrt{\Delta}}), \quad f \in L_\infty(\mathbb{R}^d), \quad g \in L_\infty(\mathbb{S}^{d-1}).$$

Let $\mathcal{A}_1 = \mathbb{C} + C_0(\mathbb{R}^d)$ and $\mathcal{A}_2 = C(\mathbb{S}^{d-1})$. Let Π be the C^*-subalgebra in $B(L_2(\mathbb{R}^d))$ generated by the algebras $\pi_1(\mathcal{A}_1)$ and $\pi_2(\mathcal{A}_2)$.

According to [59], there exists an $*$-homomorphism

$$\text{sym} : \Pi \to \mathcal{A}_1 \otimes_{\min} \mathcal{A}_2 \simeq C(\mathbb{S}^{d-1}, \mathbb{C} + C_0(\mathbb{R}^d)) \tag{1.1.1}$$

such that

$$\text{sym}(\pi_1(f)) = f \otimes 1, \quad \text{sym}(\pi_2(g)) = 1 \otimes g.$$

Here, $\mathcal{A}_1 \otimes_{\min} \mathcal{A}_2$ is the minimal tensor product of the C^*-algebras $\mathcal{A}_1$ and $\mathcal{A}_2$ (see Propositions 1.22.2 and 1.22.3 in [58]). Elements of $\mathcal{A}_1 \otimes_{\min} \mathcal{A}_2$ are identified with continuous functions on $\mathbb{R}^d \times \mathbb{S}^{d-1}$. This $*$-homomorphism is called a principal symbol mapping. It properly extends the notion of the principal symbol of a classical pseudodifferential operator (see Lemmas 8.1 and 8.2 in [59]).

It is natural to ask whether this C^*-algebraic approach works in the general setting of smooth compact manifolds. Higson (in private communication) suggested to de-manifoldise the question and reformulate it in a purely Euclidean fashion. Motivated by his suggestion, we state the natural question on the properties of the C^*-algebra Π.

Question 1.1.2 The natural unitary action of the group of diffeomorphisms on $\mathbb{R}^d$ is defined as follows. Let $\Phi : \mathbb{R}^d \to \mathbb{R}^d$ be a diffeomorphism. Let $U_\Phi \in B(L_2(\mathbb{R}^d))$ be a unitary operator given by setting

$$U_\Phi \xi = |\det(J_\Phi)|^{\frac{1}{2}} \cdot (\xi \circ \Phi), \quad \xi \in L_2(\mathbb{R}^d).$$

Here, J_Φ is the Jacobian matrix of Φ.

Is the C^*-algebra Π invariant under the action $T \to U_\Phi^{-1} T U_\Phi$? Does the $*$-homomorphism sym behave equivariantly under this action?

A *positive* answer to Question 1.1.2 (under the additional requirement that Φ is affine outside of some ball) is given in [43]. This additional assumption yields, in particular, that Φ extends to a diffeomorphism of the projective space $P^d(\mathbb{R})$. Moreover, the paper [43] establishes invariance of Π and equivariance of sym under local diffeomorphisms.

Chapter 1
Introduction

1.1 C^*-Algebraic Approach to the Principal Symbol Mapping on Smooth Manifolds

A central notion of the theory of pseudodifferential operators (PSDOs) is that of a principal symbol, which is roughly a homomorphism from the algebra of pseudo-differential operators into an algebra of functions, [41, Lemma 5.1], [39, Theorem 5.5], [63, pp. 54–55]. Usually, it is defined in a manner inhospitable for operator theorists. However, in [59], a new approach (in the setting of PSDOs on $\mathbb{R}^d$) to a principal symbol mapping on a certain C^*-subalgebra Π in $B(L_2(\mathbb{R}^d))$ is proposed; this mapping turns out to be a $*$-homomorphism from Π into a commutative C^*-algebra. The C^*-algebra Π contains all classical compactly based pseudodifferential operators on $\mathbb{R}^d$. This provides a C^*-algebraic approach to the theory which is arguably more "user friendly". The latter theory was extended to the setting of smooth compact manifolds in [43].

Whereas the approach in [59] is more elementary than the classical approach, the C^*-algebra Π introduced in [59] (see also [47]) is much wider than the class of a classical compactly based pseudo-differential operators of order 0 on $\mathbb{R}^d$. Here comes the key point: the C^*-algebra Π is the closure (in the uniform norm) of the $*$-algebra of all compactly supported *classical* pseudodifferential operators of order 0 (on $\mathbb{R}^d$). However, we use an elementary definition of Π which does not involve pseudodifferential operators. The idea to consider this closure may be discerned yet in [3] (see Proposition 5.2 on p.512). For the recent development of this idea we refer to [47, 59].

Let us explain the definition of Π and that of the principal symbol mapping in the setting of Euclidean space $\mathbb{R}^d$. Let $D_k = \frac{\partial}{\iota \partial t_k}$ be the k-th partial derivative operator on $\mathbb{R}^d$ (these are unbounded self-adjoint operators on $L_2(\mathbb{R}^d)$ (here, $\iota = \sqrt{-1}$). Denote $\nabla = (D_1, \cdots, D_d)$ and $\Delta = \sum_{k=1}^d D_k^2$. Let the d-dimensional vector $\frac{\nabla}{\Delta^{\frac{1}{2}}}$ be defined by the functional calculus (its spectrum is the unit sphere $\mathbb{S}^{d-1}$ and its spectral measure is absolutely continuous with respect to the Lebesgue measure on the sphere). Let M_f be the multiplication operator by a function f.

Definition 1.1.1 Let $\pi_1 : L_\infty(\mathbb{R}^d) \to B(L_2(\mathbb{R}^d))$, $\pi_2 : L_\infty(\mathbb{S}^{d-1}) \to B(L_2(\mathbb{R}^d))$ be defined by setting

$$\pi_1(f) = M_f, \quad \pi_2(g) = g(\frac{\nabla}{\sqrt{\Delta}}), \quad f \in L_\infty(\mathbb{R}^d), \quad g \in L_\infty(\mathbb{S}^{d-1}).$$

Let $\mathcal{A}_1 = \mathbb{C} + C_0(\mathbb{R}^d)$ and $\mathcal{A}_2 = C(\mathbb{S}^{d-1})$. Let Π be the C^*-subalgebra in $B(L_2(\mathbb{R}^d))$ generated by the algebras $\pi_1(\mathcal{A}_1)$ and $\pi_2(\mathcal{A}_2)$.

According to [59], there exists an $*$-homomorphism

$$\text{sym} : \Pi \to \mathcal{A}_1 \otimes_{\min} \mathcal{A}_2 \simeq C(\mathbb{S}^{d-1}, \mathbb{C} + C_0(\mathbb{R}^d)) \tag{1.1.1}$$

such that

$$\text{sym}(\pi_1(f)) = f \otimes 1, \quad \text{sym}(\pi_2(g)) = 1 \otimes g.$$

Here, $\mathcal{A}_1 \otimes_{\min} \mathcal{A}_2$ is the minimal tensor product of the C^*-algebras $\mathcal{A}_1$ and $\mathcal{A}_2$ (see Propositions 1.22.2 and 1.22.3 in [58]). Elements of $\mathcal{A}_1 \otimes_{\min} \mathcal{A}_2$ are identified with continuous functions on $\mathbb{R}^d \times \mathbb{S}^{d-1}$. This $*$-homomorphism is called a principal symbol mapping. It properly extends the notion of the principal symbol of a classical pseudodifferential operator (see Lemmas 8.1 and 8.2 in [59]).

It is natural to ask whether this C^*-algebraic approach works in the general setting of smooth compact manifolds. Higson (in private communication) suggested to de-manifoldise the question and reformulate it in a purely Euclidean fashion. Motivated by his suggestion, we state the natural question on the properties of the C^*-algebra Π.

Question 1.1.2 The natural unitary action of the group of diffeomorphisms on $\mathbb{R}^d$ is defined as follows. Let $\Phi : \mathbb{R}^d \to \mathbb{R}^d$ be a diffeomorphism. Let $U_\Phi \in B(L_2(\mathbb{R}^d))$ be a unitary operator given by setting

$$U_\Phi \xi = |\det(J_\Phi)|^{\frac{1}{2}} \cdot (\xi \circ \Phi), \quad \xi \in L_2(\mathbb{R}^d).$$

Here, J_Φ is the Jacobian matrix of Φ.

Is the C^*-algebra Π invariant under the action $T \to U_\Phi^{-1} T U_\Phi$? Does the $*$-homomorphism sym behave equivariantly under this action?

A *positive* answer to Question 1.1.2 (under the additional requirement that Φ is affine outside of some ball) is given in [43]. This additional assumption yields, in particular, that Φ extends to a diffeomorphism of the projective space $P^d(\mathbb{R})$. Moreover, the paper [43] establishes invariance of Π and equivariance of sym under local diffeomorphisms.

1.1 C*-Algebraic Approach to the Principal Symbol Mapping on Smooth...

The resolution of Question 1.1.2 has opened an avenue for the definition of the C^*-algebra Π_X associated with an arbitrary compact smooth manifold X. This C^*-algebra has a remarkable property: it admits a $*$-homomorphism $\text{sym}_X : \Pi_X \to C(S^*X)$, where S^*X is the cosphere bundle of X. If $X = \mathbb{R}^d$, then sym_X coincides with the mapping sym above. Every *classical* order 0 pseudodifferential operator T on X belongs to Π_X and its principal symbol in the sense of pseudodifferential operators equals $\text{sym}_X(T)$. On the other hand, not every element of Π_X is pseudodifferential (e.g. because principal symbol of a pseudodifferential operator is necessarily smooth, while that of element of Π_X is only continuous). An approach to pseudodifferential calculi based on C^*-algebras theory was first suggested by Cordes [15] (see [49] for the case of a compact manifold).

Let us briefly describe the construction of Π_X via the patching process. Suppose X is a compact smooth manifold with an atlas $(\mathcal{U}_i, h_i)_{i \in \mathbb{I}}$. We will fix a sufficiently good measure ν on X, given by a continuous positive density. If $T \in B(L_2(X, \nu))$ is compactly supported in some chart $(\mathcal{U}_i, h_i)$ (i.e., there exists $\phi \in C_c^\infty(\mathcal{U}_i)$ such that $T = TM_\phi = M_\phi T$), then, by composing with h_i, we can transfer T to an operator on $L_2(\mathbb{R}^d)$.

Definition 1.1.3 Let X be a compact smooth manifold equipped with a continuous positive density ν and let $T \in B(L_2(X, \nu))$. We say that $T \in \Pi_X$ if

(i) for every $i \in \mathbb{I}$ and for every $\phi \in C_c(\mathcal{U}_i)$, the operator $M_\phi T M_\phi$ transferred to an operator on $L_2(\mathbb{R}^d)$ belongs to Π;
(ii) for every $\psi \in C(X)$, the operator $[T, M_\psi]$ is compact.

The following theorem is one of the main results in [43].

Theorem 1.1.4 *If X is a compact smooth manifold and if ν is a continuous positive density on X, then Π_X is a C^*-algebra and there exists a surjective $*$-homomorphism*

$$\text{sym}_X : \Pi_X \to C(S^*X)$$

such that

$$\ker(\text{sym}_X) = \mathcal{K}(L_2(X, \nu)).$$

In other words, we have a short exact sequence

$$0 \to \mathcal{K}(L_2(X, \nu)) \xrightarrow{\text{id}} \Pi_X \xrightarrow{\text{sym}_X} C(S^*X) \to 0.$$

This short exact sequence first appeared in [3] (see Proposition 5.2 on p. 512) and plays an important role in index theory (see e.g. [10, Section 24.1.8] or [5, Section 2]). It is essentially equivalent to the fact that for any operator $T \in \Pi_X$ with principal symbol $a \in C(S^*X)$,

$$\inf\{\|T + K\|_\infty : K \in \mathcal{K}(L_2(X, \nu))\} = \|a\|_{C(S^*X)}.$$

As a corollary of Theorem 1.1.4, [43] provides a version of Connes Trace Theorem (see Theorem 1.1.5 below). As stated, it extends Theorem 1 in [14]. Connes Trace Theorem is ubiquitous in Non-commutative Geometry. It serves as a ground for defining a general notion of the non-commutative integral and non-commutative Yang-Mills action (that is, Theorem 14 in [14] is taken as a definition in the non-commutative setting).

We now compare our Theorem 1.1.5 with various versions of Connes Trace Theorem available in the literature. Original proof of Connes was, according to [36] "somewhat telegraphic". For example, it was not mentioned in [14] that the pseudodifferential operator featuring in Theorem 1 in [14] is classical. Two proofs are given in [36] (Theorem 7.18 on p. 293). Despite their critique of Connes exposition, the authors of [36] also do not mention the classicality of their pseudodifferential operator. Another two proofs are given in [1]. As the authors of [1] admit, their proofs are quite sketchy, however, they provide a correct statement. The advantage of the C^*-algebraic approach in [43] is threefold: (a) a strictly larger class of operators; (b) a strictly larger class of traces; (c) a convenient category of C^*-algebras (i.e., non-commutative topological spaces), unlike the setting of classical pseudodifferential operators which does not have a natural counterpart in Non-commutative Geometry.

Theorem below is the second main result in [43].

Theorem 1.1.5 *Let φ be a normalised continuous trace on $\mathcal{L}_{1,\infty}$. Let (X, G) be a compact Riemannian manifold and let ν be the Riemannian volume. If $T \in \Pi_X$, then*

$$\varphi(T(1 + \Delta_G)^{-\frac{d}{2}}) = c_d \int_{T^*X} \mathrm{sym}_X(T) e^{-q_X} d\lambda,$$

where λ is the Liouville measure on T^*X and e^{-q_X} is the canonical weight of the Riemannian manifold.

When T is a classical pseudodifferential operator, the right hand side coincides with Wodzicki residue of $T(1 + \Delta_G)^{-\frac{d}{2}}$. We refer the reader to the extensive discussion of this matter in [44].

Note an additional assumption in Theorem 1.1.5 in comparison to Theorem 1.1.4. In the latter theorem, the (smooth compact) manifold is rather arbitrary, while in the former it is Riemannian. The Riemannian structure of X in Theorem 1.1.5 is needed in two places: (a) there is no natural measure on the cosphere bundle of an arbitrary smooth manifold (but such a measure arises naturally if the manifold is Riemannian) (b) Riemannian structure provides us with a natural second order differential operator (i.e, Laplace-Beltrami operator).

1.2 Contact Manifolds

Contact manifolds, the main object in this paper, are an important class of corank 1 sub-Riemannian manifolds that includes the Heisenberg group, strongly pseudo-convex CR structures, Sasakian manifolds and Yang-Mills type structures. They naturally appear in partial differential equations when dealing with operators with double characteristics and in complex analysis. As well known, a distinctive feature of sub-Riemannian manifolds is that the basic geometric differential operators on such manifolds are not elliptic, but only hypoelliptic. The study of such operators requires the development of new methods. For contact manifolds, the relevant pseudodifferential calculus (the Heisenberg calculus) was built independently by Beals and Greiner [7] and Taylor [61] extending previous works of Folland and Stein [34], Dynin [23, 24] and some others. The approaches of Beals and Greiner [7] and Taylor [61] are quite close, but somewhat different. They both use the idea of Folland and Stein [34] that a pseudodifferential operator on a given contact manifold at any point should be modeled in a suitable sense on a left-invariant pseudodifferential operator (a model operator) on a certain nilpotent Lie group (called the osculating group). In particular, the relevant analog of the tangent bundle in this setting should be the bundle of osculating groups. The osculating groups are isomorphic (as Lie groups) to the Heisenberg group. Generally, one can consider filtered manifolds whose osculating group may be arbitrary graded nilpotent Lie groups (stratified groups). The definition of the model left-invariant pseudodifferential operators on the osculating groups is based on homogeneous convolution operators, where homogeneity is understood with respect to some anisotropic dilation on the group. The calculus of Beals-Greiner uses special local coordinates near each point, called y-coordinates in [7] or privileged coordinates in other papers. This approach is better adapted to the case when the osculating groups may vary from point to point and holds in a more general setting of Heisenberg manifolds, which includes not only contact manifolds, but also codimension one foliations and confolations of Eliashberg and Thurston [25]. The approach of Taylor in [61] is based on the use of local coordinates given by Darboux theorem for contact manifolds. Pseudodifferential operators on Heisenberg manifolds are not standard pseudodifferential operators. They are locally pseudodifferential operators with symbols from class $S^m_{1/2,1/2}$ (see, for instance, [7, Proposition 10.22] or [61, p. 99, Eq. (1.3)]).

Later on, some other approaches to the Heisenberg calculus were suggested. The geometric approach of Epstein and Melrose [27] is based on compactifications of the cotangent bundle. It is similar to the approach of Melrose [50] in the setting of the b-calculus for manifolds with boundary. The approach of [11] in the case of Heisenberg groups is essentially equivalent to the approach of Taylor [61], but based on operator kernels, rather than symbols. Finally, in [66], van Erp and Yuncken suggested a coordinate-free definition of pseudodifferential operators based on the tangent groupoid construction and work by Debord and Skandalis [19]. The fact that the Heisenberg calculus of Beals and Greiner on a contact manifold or a codimension one foliation coincides with the groupoid calculus of [66] is observed in [66] and proved in detail in [16].

The key role in any pseudodifferential calculus is played by the notion of principal symbol. In [7, 61] the principal symbol of a Heisenberg pseudodifferential operator is defined in local coordinates only, so the definition a priori depends on the choice of these coordinates. An coordinate-free definition of the principal symbol in this setting was given first by Ponge in [54]. The principal symbol of an operator is a polyhomogeneous function with respect to the appropriate dilations on the duals of the osculating groups. Another invariant definition was given in [26, 27] as a section over a bundle of jets of vector fields representing the bundle of osculating groups.

In [64, 65], van Erp developed the C^*-algebraic approach to the principal symbol map in the Heisenberg calculus. The C^*-algebraic point of view is not useful when one is interested in regularity properties of operators, but it is convenient for the purposes of index theory. The principal symbol of an operator is defined as a section of some locally trivial bundle of C^*-algebras (or, equivalently, as an element of some groupoid C^*-algebra) associated with the bundle of osculating groups. Such a section is given by the family of the model operators on the osculating groups.

In this paper we suggest another construction of a pseudodifferential calculus and principal symbol map on compact contact manifolds. Our approach is based on the C^*-algebraic point of view and was developed in different settings in previous works [43, 47, 59]. It is closer to that of [61, 64, 65] and uses Darboux coordinates. Unlike [61, 64, 65], our approach doesn't use operator kernels and convolution operators. It directly treats the model operators as operators in a Hilbert space and elements of the von Neumann algebras of the osculating groups that allows us to apply directly tools of operator theory and harmonic analysis on the Heisenberg group. As a consequence, as in [65], we get a short exact sequence of C^*-algebras associated to the principal symbol map in the Heisenberg calculus (see Theorem 1.3.18 below) with standard applications in the index theory.

1.3 Principal Symbol Mapping on Contact Manifolds

We start our considerations with the case of the Heisenberg group. The Heisenberg group $\mathbb{H}^d$, $d \in \mathbb{N}$, is $\mathbb{C}^d \times \mathbb{R}$ equipped with the product

$$(z, t) \times (z', t') = \left(z + z', t + t' + \Im(\sum_{j=1}^{d} z_j \bar{z}'_j)\right).$$

Clearly, $\mathbb{H}^d$ is a stratified Lie group of degree 2. Its first stratum is $\mathbb{C}^d \times \{0\}$ and the second one is $\{0\} \times \mathbb{R}$.

We denote $\Re(z_l) = x_l$ and $\Im(z_l) = y_l$, $1 \le l \le d$.

1.3 Principal Symbol Mapping on Contact Manifolds

We equip $\mathbb{H}^d$ with the anisotropic dilation $\delta_r(z, t) = (rz, r^2 t)$, $r > 0$, the Haar measure $dh = dx_1 \ldots dx_d dy_1 \ldots dy_d dt$ and the Koranyi metric $d_K([z, t]) = (|z|^4 + |t|^2)^{\frac{1}{4}}$.

The $2d + 1$ vector fields

$$X_l := \frac{\partial}{\partial x_l} - y_l \frac{\partial}{\partial t}, \quad Y_l := \frac{\partial}{\partial y_l} + x_l \frac{\partial}{\partial t}, \quad 1 \leq l \leq d, \quad T := \frac{\partial}{\partial t},$$

form a natural basis for the Lie algebra of left-invariant vector fields on $\mathbb{H}^d$. For convenience, we set $X_{d+l} = Y_l$, $1 \leq l \leq d$, and $X_{2d+1} = T$.

Let $\lambda : \mathbb{H}^d \to B(L_2(\mathbb{H}^d))$ be the left regular representation of $\mathbb{H}^d$. This allows to define a representation of $C_c(\mathbb{H}^d)$ as follows. Given $f \in C_c(\mathbb{H}^d)$, the operator $\lambda(f)$ on $L_2(\mathbb{H}^d)$ is defined by

$$(\lambda(f)\xi)(h) = \int_{\mathbb{H}^d} f(h\gamma^{-1})\xi(\gamma) d\gamma, \quad \xi \in L_2(\mathbb{H}^d), \quad h \in \mathbb{H}^d.$$

The group von Neumann algebra $\mathrm{VN}(\mathbb{H}^d)$ is defined to be the closure of $\lambda(C_c(\mathbb{H}^d))$ in the weak operator topology of $B(L_2(\mathbb{H}^d))$. An operator $A \in B(L_2(\mathbb{H}^d))$ is called dilation invariant, if it commutes with the operators $\sigma_r \in B(L_2(\mathbb{H}^d))$, $r > 0$, induced by the anisotropic dilations δ_r, $r > 0$. The collection of all dilation invariant elements in $\mathrm{VN}(\mathbb{H}^d)$ is called the homogeneous group algebra and denoted by $\mathrm{VN}_{\mathrm{hom}}(\mathbb{H}^d)$. Both $\mathrm{VN}(\mathbb{H}^d)$ and $\mathrm{VN}_{\mathrm{hom}}(\mathbb{H}^d)$ are semifinite von Neumann algebras.

The standard sub-Laplacian Δ on $\mathbb{H}^d$ is a second order differential operator defined by

$$\Delta = -\sum_{l=1}^{2d} X_l^2. \tag{1.3.1}$$

For every $1 \leq k \leq 2d$, the Riesz transform R_k on $\mathbb{H}^d$ is defined by the formula $R_k = X_k \Delta^{-\frac{1}{2}}$. Denote by $C^*(\{R_k\}_{k=1}^{2d})$ the C^*-subalgebra in $B(L_2(\mathbb{H}^d))$ generated by $\{R_k\}_{k=1}^{2d}$. It is clear that R_k, $1 \leq k \leq 2d$, are dilation invariant operators and, therefore, $C^*(\{R_k\}_{k=1}^{2d})$ is a C^*-subalgebra of $\mathrm{VN}_{\mathrm{hom}}(\mathbb{H}^d)$.

The following definition should be compared with Definition 1.1.1 in the Euclidean case (originated in [59] and [47]).

Definition 1.3.1 Let $\pi_1 : L_\infty(\mathbb{H}^d) \to B(L_2(\mathbb{H}^d))$, $\pi_2 : \mathrm{VN}_{\mathrm{hom}}(\mathbb{H}^d) \to B(L_2(\mathbb{H}^d))$ be defined by setting

$$\pi_1(f) = M_f, \quad \pi_2(g) = g, \quad f \in L_\infty(\mathbb{H}^d), \quad g \in \mathrm{VN}_{\mathrm{hom}}(\mathbb{H}^d).$$

Let $\mathcal{A}_1 = (C_0 + \mathbb{C})(\mathbb{H}^d)$ and $\mathcal{A}_2 = C^*(\{R_k\}_{k=1}^{2d})$. Let Π be the C^*-subalgebra in $B(L_2(\mathbb{H}^d))$ generated by the algebras $\pi_1(\mathcal{A}_1)$ and $\pi_2(\mathcal{A}_2)$.

The following theorem arises as an application of Theorem 3.3 in [47] (see also Theorem 5.2.6 in [46]).

Theorem 1.3.2 *There exists a surjective ∗-homomorphism*

$$\text{sym} : \Pi \to \mathcal{A}_1 \otimes_{\min} \mathcal{A}_2 = (C_0 + \mathbb{C})(\mathbb{H}^d) \otimes_{\min} C^*(\{R_k\}_{k=1}^{2d}) \qquad (1.3.2)$$

such that

$$\text{sym}(\pi_1(f)) = f \otimes 1, \quad \text{sym}(\pi_2(g)) = 1 \otimes g.$$

Furthermore, $\mathcal{K}(L_2(\mathbb{H}^d)) \subset \Pi$ *and*

$$\ker(\text{sym}) = \mathcal{K}(L_2(\mathbb{H}^d)).$$

This ∗-homomorphism is called the principal symbol mapping.

In [30], the authors define an operator valued principal symbol for pseudodifferential operators on the Heisenberg group $\mathbb{H}^d$ as a measurable field of operators on $\mathbb{H}^d \times \widehat{\mathbb{H}^d}$ where $\widehat{\mathbb{H}^d}$ is the unitary dual. This definition is based on polyhomogeneous expansions and agrees with the above definition upon decomposing the algebra $\text{VN}_{\text{hom}}(\mathbb{H}^d)$ into its irreducible components.

A contact manifold is a smooth manifold X of odd dimension $2d + 1$, which is equipped with a hyperplane bundle $H \subset TX$ (i.e., a smooth sub-bundle of codimension 1) such that for an arbitrary non-vanishing local 1-form θ on X with $\theta(H) = 0$, the volume form $\theta \wedge (d\theta)^d$ is nowhere vanishing (see e.g. Appendix 4 in [2]).

A model example of a contact manifold is given by the Heisenberg group.

Example 1.3.3 Let $F : \mathbb{H}^d = \mathbb{C}^d \times \mathbb{R} \to \mathbb{R}^{2d+1}$ be a smooth mapping defined by the formula

$$F(z_1, \cdots, z_d, t) = (y_1, \cdots, y_d, -x_1, \cdots, -x_d, 1).$$

The hyperplane bundle

$$\mathcal{H} = \{\mathcal{H}_p = F(p)^\perp \subset T_p\mathbb{H}^d : p \in \mathbb{H}^d\}$$

defines a contact structure on $\mathbb{H}^d$, where we identify $T_p\mathbb{H}^d$ with $\mathbb{R}^{2d+1}$, using the structure of linear space on $\mathbb{H}^d = \mathbb{C}^d \times \mathbb{R} \cong \mathbb{R}^{2d+1}$, and $F(p)^\perp$ means the orthogonal complement of $F(p)$ with respect to the standard Euclidean structure on $\mathbb{R}^{2d+1}$. One can take

$$\theta = \sum_{j=1}^{d}(y_j dx_j - x_j dy_j) + dt.$$

It is clear that $\theta \wedge (d\theta)^d$ is non-degenerate and $\theta(\mathcal{H}) = 0$.

1.3 Principal Symbol Mapping on Contact Manifolds

We equip $\mathbb{H}^d$ with Carnot-Caratheodory metric defined as follows. Smooth curve γ is called horizontal if $\gamma'(t) \in \mathcal{H}_{\gamma(t)}$ for every t. Carnot–Caratheodory distance between points $p_1, p_2 \in \mathbb{H}^d$ is the infimum of the lengths of horizontal curves joining p_1 and p_2. The concrete formula for Carnot-Caratheodory metric is available in [8].

Every contact manifold (X, H) is locally isomorphic to an open subset of the Heisenberg group $\mathbb{H}^d$ equipped with the standard contact structure (this is a highly non-trivial result, known as Darboux theorem; see e.g. Appendix 4.H in [2]). More precisely, for every $p \in X$ there exists an open neighborhood $U \subset X$ of p and a chart $h : U \to \mathbb{H}^d$ such that dh takes the hyperplane bundle $H \subset TX$ to the hyperplane bundle $\mathcal{H} \subset T\mathbb{H}^d$ described in Example 1.3.3. This fact allows us to describe contact manifolds using Heisenberg atlases.

First, we introduce some general notation.

Notation 1.3.4 Let X be a smooth $(2d + 1)$-dimensional manifold with atlas $(\mathcal{U}_i, h_i)_{i \in \mathbb{I}}$, where $\mathbb{I}$ is an arbitrary set of indices.

(i) We denote

$$\Omega_i = h_i(\mathcal{U}_i) \subset \mathbb{R}^{2d+1}, \quad \Omega_{i,j} = h_i(\mathcal{U}_i \cap \mathcal{U}_j) \subset \mathbb{R}^{2d+1}, \quad i, j \in \mathbb{I};$$

(ii) We denote by $\Phi_{i,j} : \Omega_{i,j} \to \Omega_{j,i}$ the diffeomorphism given by the formula

$$\Phi_{i,j}(p) = h_j(h_i^{-1}(p)), \quad p \in \Omega_{i,j}.$$

Now we need the notion of Heisenberg diffeomorphism. For a diffeomorphism $\Phi : \Omega \subset \mathbb{R}^{2d+1} \to \Omega' \subset \mathbb{R}^{2d+1}$ (with components Φ_k, $1 \leq k \leq 2d+1$) we denote by $J_\Phi : \Omega \to GL(2d+1, \mathbb{R})$ the Jacobian matrix of Φ:

$$J_\Phi = (\iota D_l \Phi_k)_{k,l=1}^{2d+1},$$

where ι denotes the imaginary unit and $D_l = \frac{1}{\iota} \frac{\partial}{\partial p_l}$, $p \in \mathbb{R}^{2d+1}$.

Definition 1.3.5 A diffeomorphism $\Phi : \Omega \subset \mathbb{H}^d \cong \mathbb{R}^{2d+1} \to \Omega' \subset \mathbb{H}^d \cong \mathbb{R}^{2d+1}$ is a Heisenberg one if

$$J_\Phi(p)(F(p)^\perp) = F(\Phi(p))^\perp, \quad p \in \Omega.$$

Here we use notation introduced in Example 1.3.3.

Definition 1.3.6 A Heisenberg atlas for a contact manifold X is an atlas $(\mathcal{U}_i, h_i)_{i \in \mathbb{I}}$, where each diffeomorphism $\Phi_{i,j}$ is a Heisenberg diffeomorphism if the coordinate space $\mathbb{R}^{2d+1}$ is equipped with the canonical contact structure of the Heisenberg group $\mathbb{H}^d$.

Let $\mathfrak{B}$ be the Borel σ-algebra on X. We need the notion of density on a manifold available, e.g. in [51, p.87].

Definition 1.3.7 Let ν be a countably additive measure on $\mathfrak{B}$. We assume that, for every $i \in \mathbb{I}$, the measure $\nu \circ h_i^{-1}$ on Ω_i is absolutely continuous with respect to the Lebesgue measure on Ω_i, and its Radon-Nikodym derivative a_i is strictly positive and smooth on Ω_i.

In this case, we say that ν is a smooth positive density on X.

Now we can give the definition of the C^*-algebra Π_X, the domain of the principal symbol mapping.

Definition 1.3.8 Let X be a compact contact manifold with a Heisenberg atlas $\{(\mathcal{U}_i, h_i)\}_{i \in \mathbb{I}}$. Let ν be a smooth positive density on X. The domain Π_X of the principal symbol mapping consists of operators $S \in B(L_2(X, \nu))$ such that

(i) for every $i \in \mathbb{I}$ and for every $\phi \in C_c(\mathcal{U}_i)$, the operator $M_\phi S M_\phi$ transferred to an operator on $L_2(\mathbb{H}^d)$ belongs to Π;
(ii) for every $\psi \in C(X)$, the operator $[S, M_\psi]$ is compact.

The codomain of the principal symbol mapping is defined as the C^*-algebra of continuous sections of some locally trivial bundle of C^*-algebras on X. First, we recall general definitions.

Definition 1.3.9 Let X be a smooth manifold with an atlas $\{(\mathcal{U}_i, h_i)\}_{i \in \mathbb{I}}$. Let $\mathcal{M}$ be a unital C^*-algebra. Let $\varpi = \{\pi_{i,j}\}_{i,j \in \mathbb{I}}$ be a family of continuous mappings $\pi_{i,j} : \mathcal{U}_i \cap \mathcal{U}_j \to \mathrm{Aut}(\mathcal{M})$ defined on each nonempty overlap $\mathcal{U}_i \cap \mathcal{U}_j$ such that the following conditions hold:

$$\begin{aligned} \pi_{i,j}(x)\pi_{j,i}(x) &= \mathrm{id}, \quad x \in \mathcal{U}_i \cap \mathcal{U}_j, \\ \pi_{i,j}(x)\pi_{j,k}(x)\pi_{k,i}(x) &= \mathrm{id}, \quad x \in \mathcal{U}_i \cap \mathcal{U}_j \cap \mathcal{U}_k. \end{aligned} \quad (1.3.3)$$

The triple $(X, \mathcal{M}, \varpi)$ is called a bundle of C^*-algebras. More precisely, it is a locally trivial bundle with fiber $\mathcal{M}$ and structure group $\mathrm{Aut}(\mathcal{M})$ that is trivializable over $\{\mathcal{U}_i\}$ with transition functions $\{\pi_{i,j}\}_{i,j \in \mathbb{I}}$.

Definition 1.3.10 The space $C(E)$ of bounded continuous sections of a bundle of C^*-algebras $E = (X, \mathcal{M}, \varpi)$ is the set of all families $F = \{F_i\}_{i \in \mathbb{I}}$ such that

(i) for every $i \in \mathbb{I}$, we have $F_i \in C(\mathcal{U}_i, \mathcal{M})$;
(ii) for every $i, j \in \mathbb{I}$, we have $F_i = \pi_{i,j}(F_j)$ on $\mathcal{U}_i \cap \mathcal{U}_j$;
(iii)

$$\|F\|_{C(E)} := \sup_{i \in \mathbb{I}} \|F_i\|_{C(\mathcal{U}_i, \mathcal{M})} < \infty.$$

The following result is elementary.

Theorem 1.3.11 *If E is a bundle of C^*-algebras, then $C(E)$ is a C^*-algebra.*

1.3 Principal Symbol Mapping on Contact Manifolds

Just like $C(E)$ is obtained by gluing the algebras $C(\mathcal{U}_i, \mathcal{M})$ of bounded continuous $\mathcal{M}$-valued functions, the algebra $L_\infty(E)$ is obtained by gluing algebras $L_\infty(\mathcal{U}_i, \mathcal{M})$ of bounded ultraweakly measurable $\mathcal{M}$-valued functions.

Definition 1.3.12 Let $E = (X, \mathcal{M}, \varpi)$ be a bundle of C^*-algebras. Suppose in addition that $\mathcal{M}$ is a von Neumann algebra. The space $L_\infty(E)$ of bounded measurable sections of this bundle is the set of all families $F = \{F_i\}_{i \in \mathbb{I}}$ such that

(i) for every $i \in \mathbb{I}$, we have $F_i \in L_\infty(\mathcal{U}_i, \mathcal{M})$;
(ii) for every $i, j \in \mathbb{I}$, we have $F_i = \pi_{i,j}(F_j)$ on $\mathcal{U}_i \cap \mathcal{U}_j$;
(iii)
$$\|F\|_{L_\infty(E)} := \sup_{i \in \mathbb{I}} \|F_i\|_{L_\infty(\mathcal{U}_i, \mathcal{M})} < \infty.$$

The following result is elementary.

Theorem 1.3.13 Let $E = (X, \mathcal{M}, \varpi)$ be a bundle of C^*-algebras. Suppose in addition that $\mathcal{M}$ is a von Neumann algebra. Then $L_\infty(E)$ is a von Neumann algebra.

Let $E = (X, \mathcal{M}, \varpi)$ be a bundle of C^*-algebras. For every $i \in \mathbb{I}$, there is a natural embedding $\Xi_i : C_c(\mathcal{U}_i, \mathcal{M}) \to C(E)$. Given $f \in C_c(\mathcal{U}_i, \mathcal{M})$, the corresponding family $F = \{F_j\}_{j \in \mathbb{I}}$ is defined as follows. The function $F_j \in C(\mathcal{U}_j, \mathcal{M})$ is the extension of $\pi_{j,i}(f|_{\mathcal{U}_i \cap \mathcal{U}_j})$ to $\mathcal{U}_j$ by zero if $\mathcal{U}_i \cap \mathcal{U}_j \neq \emptyset$ and equals 0 if $\mathcal{U}_i \cap \mathcal{U}_j = \emptyset$. It is easy to see that each F_j is well-defined and continuous and the family $F = \{F_j\}_{j \in \mathbb{I}}$ satisfies the compatibility conditions.

Let us turn to the case of a compact contact manifold X. The transition functions of our C^*-algebra bundles are constructed from the transition functions of the fixed Heisenberg atlas on X. Therefore, we start our considerations with an arbitrary Heisenberg diffeomorphism.

Definition 1.3.14 Let $\Phi : \Omega \subset \mathbb{H}^d \to \Omega' \subset \mathbb{H}^d$ be a Heisenberg diffeomorphism. Define the horizontal Jacobian matrix $HJ_\Phi : \Omega \to \mathrm{GL}(2d, \mathbb{R})$ of Φ by setting

$$HJ_\Phi = (X_l \Phi_k)_{k,l=1}^{2d}.$$

The terminology "horizontal Jacobian matrix" is common in sub-Riemannian geometry; it originates from horizontal spaces for connections on fiber bundles.

Just like the Jacobian, the horizontal Jacobian satisfies the composition rule

$$HJ_{\Phi_1 \circ \Phi_2} = (HJ_{\Phi_1} \circ \Phi_2) \cdot HJ_{\Phi_2} \tag{1.3.4}$$

whenever Φ_1 and Φ_2 are Heisenberg diffeomorphisms.

For $k, l = 1, \cdots, 2d$, let $E_{k,l} \in M_{2d}(\mathbb{R})$ be the matrix whose (i, j)-entry equals 1 if $(i, j) = (k, l)$ and 0 otherwise. Let

$$\Omega = \sum_{k=1}^{d} E_{k+d,k} - E_{k,k+d} \in M_{2d}(\mathbb{R}). \tag{1.3.5}$$

Recall the following well-known fact (see e.g. Section 2.3 in [42]).

Theorem 1.3.15 *Let $\Phi : \Omega \subset \mathbb{H}^d \to \Omega' \subset \mathbb{H}^d$ be a Heisenberg diffeomorphism. There exists a smooth function λ_Φ on Ω such that:*

(i) *for every $p \in \Omega$, we have*

$$H J_\Phi(p) \cdot \Omega \cdot H J_\Phi^*(p) = \lambda_\Phi(p) \Omega.$$

(ii) *for every $p \in \Omega$, the linear mapping $H J_\Phi(p) \in GL(2d, \mathbb{R})$ produces an element $A^\Phi(p) \in \text{Aut}(\mathbb{H}^d)$ by the formula*

$$A^\Phi(p) = \begin{pmatrix} H J_\Phi(p) & 0 \\ 0 & \lambda_\Phi(p) \end{pmatrix}.$$

It is easy to see that the mappings A^Φ also satisfy the composition rule

$$A^{\Phi_1 \circ \Phi_2} = (A^{\Phi_1} \circ \Phi_2) \cdot A^{\Phi_2} \tag{1.3.6}$$

whenever Φ_1 and Φ_2 are Heisenberg diffeomorphisms.

Every $A \in \text{Aut}(\mathbb{H}^d)$ induces a $*$-automorphism π_A of $\text{VN}(\mathbb{H}^d)$ by the formula

$$\pi_A(x) = V_A^{-1} x V_A, \quad x \in \text{VN}(\mathbb{H}^d), \tag{1.3.7}$$

where

$$V_A \xi = \xi \circ A^*, \quad \xi \in L_2(\mathbb{H}^d).$$

Note the composition rule

$$\pi_A \circ \pi_B = \pi_{BA}, \quad A, B \in \text{Aut}(\mathbb{H}^d). \tag{1.3.8}$$

The $*$-automorphism π_A of $\text{VN}(\mathbb{H}^d)$ maps $C^*(\{R_k\}_{k=1}^{2d})$ into itself (see Theorem 3.1.1).

1.3 Principal Symbol Mapping on Contact Manifolds

Definition 1.3.16 Let $A : \Omega \to \mathrm{Aut}(\mathbb{H}^d)$ be a measurable function defined on an open subset $\Omega \subset \mathbb{H}^d$. The $*$-automorphism π_A of $L_\infty(\Omega) \bar{\otimes} \mathrm{VN}(\mathbb{H}^d)$ is given by the formula

$$(\pi_A(x))(p) = \pi_{A(p)}(x(p)), \quad p \in \Omega.$$

As shown in Theorem 3.1.2, if $A : \Omega \to \mathrm{Aut}(\mathbb{H}^d)$ is a continuous function defined on an open subset $\Omega \subset \mathbb{H}^d$, then the $*$-automorphism π_A maps $C(\Omega) \otimes_{\min} C^*(\{R_k\}_{k=1}^{2d})$ into itself.

Now we are ready to describe the image of the principal symbol mapping. Let X be a compact contact manifold with a Heisenberg atlas $\{(\mathcal{U}_i, h_i)\}_{i \in \mathbb{I}}$. For any $i, j \in \mathbb{I}$ such that $\mathcal{U}_i \cap \mathcal{U}_j \neq \varnothing$, the diffeomorphism

$$\Phi_{i,j} : \Omega_{i,j} = h_i(\mathcal{U}_i \cap \mathcal{U}_j) \subset \mathbb{R}^{2d+1} \to \Omega_{j,i} = h_j(\mathcal{U}_i \cap \mathcal{U}_j) \subset \mathbb{R}^{2d+1}$$

introduced in Notation 1.3.4 is a Heisenberg diffeomorphism. One can check (see Lemma 4.1.6 below) that the family $\{A^{\Phi_{i,j} \circ h_i} : \mathcal{U}_i \cap \mathcal{U}_j \to \mathrm{Aut}(\mathbb{H}^d), i, j \in \mathbb{I}\}$ as well as

$$\varpi = \{\pi_{i,j} = \pi_{A^{\Phi_{i,j} \circ h_i}} : \mathcal{U}_i \cap \mathcal{U}_j \to \mathrm{Aut}(C^*(\{R_k\}_{k=1}^{2d})), i, j \in \mathbb{I}\},$$

satisfy the conditions (1.3.3). Denote by $E_{\mathrm{hom}} = (X, \mathcal{M}, \varpi)$ the corresponding bundle of C^*-algebras.

Definition 1.3.17 The codomain of the principal symbol mapping is the algebra $C(E_{\mathrm{hom}})$ of continuous sections of E_{hom}.

The first main result of the paper is the following theorem. It is a Heisenberg analogue of Theorem 1.4 in [43] (see Theorem 1.1.4 above).

Theorem 1.3.18 *Let X be a compact contact manifold with a Heisenberg atlas $\{(\mathcal{U}_i, h_i)\}_{i \in \mathbb{I}}$. Let ν be a smooth positive density on X.*

(i) *Π_X is a C^*-algebra.*
(ii) *There exists a surjective $*$-homomorphism $\mathrm{sym}_X : \Pi_X \to C(E_{\mathrm{hom}})$ such that*

$$\ker(\mathrm{sym}_X) = \mathcal{K}(L_2(X, \nu)).$$

The global principal symbol mapping sym_X is compatible with the local one sym introduced in Theorem 1.3.2 in the following sense. If $S \in \Pi_X$ is compactly supported in $\mathcal{U}_i$ for some $i \in \mathbb{I}$, then the operator S transferred to an operator on $L_2(\mathbb{H}^d)$ belongs to Π and

$$\mathrm{sym}_X(S) = \Xi_i(\mathrm{sym}(S)).$$

1.4 Connes Trace Theorem in the Setting of Contact Manifolds

Next, we turn to the analogue of the Connes trace theorem for contact manifolds.

As above, we consider first the case of the Heisenberg group. On the subalgebra $\lambda(C_c(\mathbb{H}^d)) \subset B(L_2(\mathbb{H}^d))$, we define a functional

$$\tau(\lambda(f)) = f(0), \quad f \in C_c(\mathbb{H}^d).$$

The functional τ uniquely extends to a semifinite normal trace on the group von Neumann algebra $\mathrm{VN}(\mathbb{H}^d)$ (see Section 7.2 and, in particular, Theorem 7.2.7 in [52]).

Theorem 1.4.1 below is a Heisenberg analogue of the Euclidean Connes trace formula (see e.g. Chapter 5 in [46]). We refer the reader to [45] for the general theory of traces and to [46] for detailed exposition of the Connes trace formula.

Theorem 1.4.1 *Let φ be a normalised continuous trace on $\mathcal{L}_{1,\infty}$. Let $S \in \Pi$ be compactly supported and let Δ be the standard sub-Laplacian (see (1.3.1)). We have*

$$\varphi(S(1+\Delta)^{-d-1}) = c_d \Big(\int_{\mathbb{H}^d} \otimes \tau \Big) \Big(\mathrm{sym}(S) \cdot (1 \otimes e^{-\Delta}) \Big).$$

The right-hand side of the last formula should be understood as follows. Recall that

$$\mathrm{sym}(S) \in (C_0 + \mathbb{C})(\mathbb{H}^d) \otimes_{\min} C^*(\{R_k\}_{k=1}^{2d}) \subset L_\infty(\mathbb{H}^d) \bar{\otimes} \mathrm{VN}(\mathbb{H}^d).$$

One can show that $e^{-\Delta}$ belongs to $\mathrm{VN}(\mathbb{H}^d)$. So the product $\mathrm{sym}(S) \cdot (1 \otimes e^{-\Delta})$ is a well-defined element of $L_\infty(\mathbb{H}^d) \bar{\otimes} \mathrm{VN}(\mathbb{H}^d)$. The trace $\int_{\mathbb{H}^d} \otimes \tau$ is the tensor product of the trace $\int_{\mathbb{H}^d}$ on $L_\infty(\mathbb{H}^d)$ given by the integration over $\mathbb{H}^d$ and the trace τ on $\mathrm{VN}(\mathbb{H}^d)$. One can show that the right-hand side of the last formula is finite.

In Sect. 2.6, we extend Theorem 1.4.1 to an arbitrary elliptic, self-adjoint and positive differential operator (in Heisenberg sense) of order 2 (see Theorem 2.6.1).

Let (X, H) be a compact contact manifold with a Heisenberg atlas $\{(\mathcal{U}_i, h_i)\}_{i \in \mathbb{I}}$. Let G be a sub-Riemannian structure on X, that is, a smooth scalar product in the fibers of H. For any $i \in \mathbb{I}$, the matrix representation of the metric G in the chart $(\mathcal{U}_i, h_i)$ with respect to the basis $(X_1, \cdots, X_{2d})$ gives rise to a smooth mapping $G_i : \mathcal{U}_i \to \mathrm{GL}^+(2d, \mathbb{R})$ (in what follows, $\mathrm{GL}^+(2d, \mathbb{R})$ stands for the set of all positive elements in $\mathrm{GL}(2d, \mathbb{R})$). We also need the notation $g_i = G_i \circ h_i^{-1} : \Omega_i \to \mathrm{GL}^+(2d, \mathbb{R})$, $i \in \mathbb{I}$.

Notation 1.4.2 Let $\Omega \subset \mathbb{H}^d$ be connected and open. Let $g : \Omega \to \mathrm{GL}^+(2d, \mathbb{R})$ be a smooth mapping and ν be a smooth positive density on Ω. The sub-Laplacian

1.3 Principal Symbol Mapping on Contact Manifolds

Definition 1.3.16 Let $A : \Omega \to \mathrm{Aut}(\mathbb{H}^d)$ be a measurable function defined on an open subset $\Omega \subset \mathbb{H}^d$. The $*$-automorphism π_A of $L_\infty(\Omega)\bar\otimes \mathrm{VN}(\mathbb{H}^d)$ is given by the formula

$$(\pi_A(x))(p) = \pi_{A(p)}(x(p)), \quad p \in \Omega.$$

As shown in Theorem 3.1.2, if $A : \Omega \to \mathrm{Aut}(\mathbb{H}^d)$ is a continuous function defined on an open subset $\Omega \subset \mathbb{H}^d$, then the $*$-automorphism π_A maps $C(\Omega) \otimes_{\min} C^*(\{R_k\}_{k=1}^{2d})$ into itself.

Now we are ready to describe the image of the principal symbol mapping. Let X be a compact contact manifold with a Heisenberg atlas $\{(\mathcal{U}_i, h_i)\}_{i\in \mathbb{I}}$. For any $i, j \in \mathbb{I}$ such that $\mathcal{U}_i \cap \mathcal{U}_j \neq \emptyset$, the diffeomorphism

$$\Phi_{i,j} : \Omega_{i,j} = h_i(\mathcal{U}_i \cap \mathcal{U}_j) \subset \mathbb{R}^{2d+1} \to \Omega_{j,i} = h_j(\mathcal{U}_i \cap \mathcal{U}_j) \subset \mathbb{R}^{2d+1}$$

introduced in Notation 1.3.4 is a Heisenberg diffeomorphism. One can check (see Lemma 4.1.6 below) that the family $\{A^{\Phi_{i,j}\circ h_i} : \mathcal{U}_i \cap \mathcal{U}_j \to \mathrm{Aut}(\mathbb{H}^d), i, j \in \mathbb{I}\}$ as well as

$$\varpi = \{\pi_{i,j} = \pi_{A^{\Phi_{i,j}\circ h_i}} : \mathcal{U}_i \cap \mathcal{U}_j \to \mathrm{Aut}(C^*(\{R_k\}_{k=1}^{2d})), i, j \in \mathbb{I}\},$$

satisfy the conditions (1.3.3). Denote by $E_{\mathrm{hom}} = (X, \mathcal{M}, \varpi)$ the corresponding bundle of C^*-algebras.

Definition 1.3.17 The codomain of the principal symbol mapping is the algebra $C(E_{\mathrm{hom}})$ of continuous sections of E_{hom}.

The first main result of the paper is the following theorem. It is a Heisenberg analogue of Theorem 1.4 in [43] (see Theorem 1.1.4 above).

Theorem 1.3.18 *Let X be a compact contact manifold with a Heisenberg atlas $\{(\mathcal{U}_i, h_i)\}_{i\in \mathbb{I}}$. Let ν be a smooth positive density on X.*

(i) *Π_X is a C^*-algebra.*
(ii) *There exists a surjective $*$-homomorphism $\mathrm{sym}_X : \Pi_X \to C(E_{\mathrm{hom}})$ such that*

$$\ker(\mathrm{sym}_X) = \mathcal{K}(L_2(X, \nu)).$$

The global principal symbol mapping sym_X is compatible with the local one sym introduced in Theorem 1.3.2 in the following sense. If $S \in \Pi_X$ is compactly supported in $\mathcal{U}_i$ for some $i \in \mathbb{I}$, then the operator S transferred to an operator on $L_2(\mathbb{H}^d)$ belongs to Π and

$$\mathrm{sym}_X(S) = \Xi_i(\mathrm{sym}(S)).$$

1.4 Connes Trace Theorem in the Setting of Contact Manifolds

Next, we turn to the analogue of the Connes trace theorem for contact manifolds.

As above, we consider first the case of the Heisenberg group. On the subalgebra $\lambda(C_c(\mathbb{H}^d)) \subset B(L_2(\mathbb{H}^d))$, we define a functional

$$\tau(\lambda(f)) = f(0), \quad f \in C_c(\mathbb{H}^d).$$

The functional τ uniquely extends to a semifinite normal trace on the group von Neumann algebra $\mathrm{VN}(\mathbb{H}^d)$ (see Section 7.2 and, in particular, Theorem 7.2.7 in [52]).

Theorem 1.4.1 below is a Heisenberg analogue of the Euclidean Connes trace formula (see e.g. Chapter 5 in [46]). We refer the reader to [45] for the general theory of traces and to [46] for detailed exposition of the Connes trace formula.

Theorem 1.4.1 *Let φ be a normalised continuous trace on $\mathcal{L}_{1,\infty}$. Let $S \in \Pi$ be compactly supported and let Δ be the standard sub-Laplacian (see (1.3.1)). We have*

$$\varphi(S(1+\Delta)^{-d-1}) = c_d \Big(\int_{\mathbb{H}^d} \otimes \tau \Big) \Big(\mathrm{sym}(S) \cdot (1 \otimes e^{-\Delta}) \Big).$$

The right-hand side of the last formula should be understood as follows. Recall that

$$\mathrm{sym}(S) \in (C_0 + \mathbb{C})(\mathbb{H}^d) \otimes_{\min} C^*(\{R_k\}_{k=1}^{2d}) \subset L_\infty(\mathbb{H}^d) \bar\otimes \mathrm{VN}(\mathbb{H}^d).$$

One can show that $e^{-\Delta}$ belongs to $\mathrm{VN}(\mathbb{H}^d)$. So the product $\mathrm{sym}(S) \cdot (1 \otimes e^{-\Delta})$ is a well-defined element of $L_\infty(\mathbb{H}^d) \bar\otimes \mathrm{VN}(\mathbb{H}^d)$. The trace $\int_{\mathbb{H}^d} \otimes \tau$ is the tensor product of the trace $\int_{\mathbb{H}^d}$ on $L_\infty(\mathbb{H}^d)$ given by the integration over $\mathbb{H}^d$ and the trace τ on $\mathrm{VN}(\mathbb{H}^d)$. One can show that the right-hand side of the last formula is finite.

In Sect. 2.6, we extend Theorem 1.4.1 to an arbitrary elliptic, self-adjoint and positive differential operator (in Heisenberg sense) of order 2 (see Theorem 2.6.1).

Let (X, H) be a compact contact manifold with a Heisenberg atlas $\{(\mathcal{U}_i, h_i)\}_{i \in \mathbb{I}}$. Let G be a sub-Riemannian structure on X, that is, a smooth scalar product in the fibers of H. For any $i \in \mathbb{I}$, the matrix representation of the metric G in the chart $(\mathcal{U}_i, h_i)$ with respect to the basis $(X_1, \cdots, X_{2d})$ gives rise to a smooth mapping $G_i : \mathcal{U}_i \to \mathrm{GL}^+(2d, \mathbb{R})$ (in what follows, $\mathrm{GL}^+(2d, \mathbb{R})$ stands for the set of all positive elements in $\mathrm{GL}(2d, \mathbb{R})$). We also need the notation $g_i = G_i \circ h_i^{-1} : \Omega_i \to \mathrm{GL}^+(2d, \mathbb{R})$, $i \in \mathbb{I}$.

Notation 1.4.2 Let $\Omega \subset \mathbb{H}^d$ be connected and open. Let $g : \Omega \to \mathrm{GL}^+(2d, \mathbb{R})$ be a smooth mapping and ν be a smooth positive density on Ω. The sub-Laplacian

1.4 Connes Trace Theorem in the Setting of Contact Manifolds

$\Delta_{g,\nu} : C_c^\infty(\Omega) \to C_c^\infty(\Omega)$ is a second order differential operator defined by

$$\Delta_{g,\nu} = \sum_{k,l=1}^{2d} X_k^* M_{(g^{-1})_{k,l}} X_l,$$

where X_k^* is the formal adjoint of X_k in $L_2(\Omega, \nu)$.

One can show (see Theorem 4.2.3 and Definition 4.2.5 below) that there exists a unique second order differential operator $\Delta_{G,\nu} : C^\infty(X) \to C^\infty(X)$ such that

$$(\Delta_{G,\nu} f) \circ h_i^{-1} = \Delta_{g_i, \nu \circ h_i^{-1}}(f \circ h_i^{-1}), \quad f \in C_c^\infty(\mathcal{U}_i).$$

This operator called the sub-Laplacian is analogous to the Laplace-Beltrami operator in the Riemannian geometry.

Theorem 1.4.3 *Let (X, G) be a compact contact sub-Riemannian manifold and let ν be a smooth positive density on X. The sub-Laplacian admits a unique self-adjoint extension $\Delta_{G,\nu}$ (with domain $W^{2,2}(X)$) in $L_2(X, \nu)$.*

Theorem 1.4.3 is established in Sect. 4.2.

Consider the unital C^*-algebra $\mathrm{VN}(\mathbb{H}^d)$ and the family $\varpi = \{\pi_{i,j}\}_{i,j \in \mathbb{I}}$ of continuous mappings $\pi_{i,j} : \mathcal{U}_i \cap \mathcal{U}_j \to \mathrm{Aut}(\mathrm{VN}(\mathbb{H}^d))$ defined on each nonempty overlap $\mathcal{U}_i \cap \mathcal{U}_j$ by

$$\pi_{i,j} = \pi_{A^{\Phi_{i,j}} \circ h_i}.$$

Let $E = (X, \mathrm{VN}(\mathbb{H}^d), \varpi)$ be the associated C^*-algebra bundle given by Definition 1.3.9. Recall that $L_\infty(E)$ denotes the von Neumann algebra of bounded measurable sections of E (see Definition 1.3.12). The bundle E can be considered as an analogue of the cotangent bundle in the setting of contact manifold. The following theorem defines an analogue of the Liouville measure in this setting, being a trace on the von Neumann algebra of its L_∞-sections.

Theorem 1.4.4 *There exists a (faithful normal) semifinite trace Λ on $L_\infty(E)$ such that*

$$\Lambda(F) = \int_{\Omega_i} \tau(F_i(h_i^{-1}(p))) dp, \qquad (1.4.1)$$

for every $i \in \mathbb{I}$ and for every $F \in L_\infty(E)$ supported on $\mathcal{U}_i$.

Next we should define analogues of the cosphere bundle and Liouville measure on it. It clear that an analogue of the cosphere bundle is given by the C^*-algebra bundle $E_{\mathrm{hom}} = (X, \mathrm{VN}_{\mathrm{hom}}(\mathbb{H}^d), \varpi)$. To define an analogue of the Liouville measure on the cosphere bundle, as in [43], we will use the notion of canonical weight.

Consider the Hilbert space $L_2(\mathbb{H}^d)$ and the family $\upsilon = \{\upsilon_{i,j}\}_{i,j\in\mathbb{I}}$ of continuous mappings $\upsilon_{i,j} : \mathcal{U}_i \cap \mathcal{U}_j \to U(L_2(\mathbb{H}^d))$ defined on each nonempty overlap $\mathcal{U}_i \cap \mathcal{U}_j$ by

$$\upsilon_{i,j} = V_{A^{\Phi_{i,j}}\circ h_i}.$$

We get a locally trivial bundle of Hilbert spaces $\mathcal{H} = (X, L_2(\mathbb{H}^d), \upsilon)$ on X with fiber $L_2(\mathbb{H}^d)$ and structure group $U(L_2(\mathbb{H}^d))$. Denote by $L_2(X, \mathcal{H}, \nu)$ the Hilbert space of L_2-sections of this Hilbert bundle. There is a natural faithful $*$-representation of the von Neumann algebra $L_\infty(E)$ on $L_2(X, \mathcal{H}, \nu)$. We refer the reader to Sect. 4.4 for more details.

Lemma 1.4.5 *There exists an unbounded self-adjoint positive operator q_X on $L_2(X, \mathcal{H}, \nu)$ affiliated with $L_\infty(E)$ such that for every $i \in \mathbb{I}$, the restriction $(q_X)_i$ of q_X to $L_2(\mathcal{U}_i, \mathcal{H}, \nu) = L_2(\mathcal{U}_i, \nu) \otimes L_2(\mathbb{H}^d)$ is of the form*

$$(q_X)_i = -\sum_{k_1,k_2=1}^{2d} (G_i^{-1})_{k_1,k_2} \otimes X_{k_1} X_{k_2}.$$

By the functional calculus, we get a well-defined operator e^{-q_X} on $L_2(X, \mathcal{H}, \nu)$, which belongs to the von Neumann algebra $L_\infty(E)$.

We can now state the second main result in our paper. It is a Heisenberg analogue of Theorem 1.5 in [43] (see Theorem 1.1.5 above).

Theorem 1.4.6 *Let φ be a normalised continuous trace on $\mathcal{L}_{1,\infty}$. Let (X, G) be a compact contact sub-Riemannian manifold with a Heisenberg atlas $\{(\mathcal{U}_i, h_i)\}_{i\in\mathbb{I}}$. Let ν be a smooth positive density on X.*

(i) *We have $(1 + \Delta_{G,\nu})^{-1} \in \mathcal{L}_{d+1,\infty}$.*
(ii) *For every $S \in \Pi_X$, we have*

$$\varphi(S(1 + \Delta_{G,\nu})^{-d-1}) = c_d \Lambda(\mathrm{sym}_X(S)e^{-q_X}). \qquad (1.4.2)$$

As widely believed, this theorem is roughly equivalent to the microlocal Weyl law. For contact closed manifolds, Taylor [62] established the microlocal Weyl law and the asymptotic

$$\mathrm{Tr}(Se^{-t\Delta_{G,\nu}}) = c(S)t^{-d-1} + O(t^{-d-\frac{1}{2}}), \quad t \downarrow 0$$

for S from Taylor's algebra $\mathrm{OP}\tilde{\Psi}^0(X)$ of classical Heisenberg-PSDOs. The latter asymptotic allows to deduce (1.4.2) for $S \in \mathrm{OP}\tilde{\Psi}^0(X)$. However, our algebra Π_X is much larger than $\mathrm{OP}\tilde{\Psi}^0(X)$ (in particular, it contains all compact operators on

1.4 Connes Trace Theorem in the Setting of Contact Manifolds

$L_2(X, \nu)$). Thus, the best one can expect for $S \in \Pi_X$ is the asymptotic

$$\text{Tr}(Se^{-t\Delta_{G,\nu}}) = c(S)t^{-d-1} + o(t^{-d-1}), \quad t \downarrow 0.$$

The latter estimate is insufficient to derive Theorem 1.4.6 in its full strength. Our method of proving Theorem 1.4.6 is, therefore, very different from that of Taylor and does not rely on the asymptotic for the heat semi-group.

For $S \in OPS^0(M)$, the algebra of classical pseudodifferential operators, such an asymptotic was proved by Colin de Verdière, Hillairet, and Trélat in [12] (see also [13]). The noncommutative residue in the Beals-Greiner calculus on Heisenberg manifolds was introduced by Ponge in [53]. In [18], Dave and Haller defined and studied the noncommutative residue for filtered manifolds generalizing that of Ponge for Heisenberg manifolds. In the framework of the groupoidal calculus on Heisenberg manifolds developed in [66], the noncommutative residue was studied by Couchet and Yuncken in [17]. In particular, it was shown that such a groupoidal residue coincides with Ponge's definition [53] for contact manifolds and codimension one foliations. It looks likely that the right hand side of (1.4.2) agrees with the noncommutative residue defined in [17, 18, 53] but we do not pursue the question here. In [53, Theorem 4.12], Ponge proved that, for a classical pseudodifferential operators in the Beals-Greiner calculus of order less than or equal to $-(\dim M + 1)$, the Dixmier trace is well defined and agrees with the noncommutative residue. This result is the analogue of the original Connes' trace theorem for classical pseudodifferential operators on a compact manifold. As a consequence, the author derives the Connes' trace theorem for the sub-Laplacian on a contact compact manifold associated with the metric, compatible with the contact structure. Our result, Theorem 1.4.6, is more general. It deals with much larger algebra of operators and with an arbitrary normalised continuous trace on $\mathcal{L}_{1,\infty}$.

As a corollary, we obtain a spectrally correct sub-Riemannian volume vol_G on (X, G) given by

$$\int_X f d\text{vol}_G = \Lambda(fe^{-q_X}), \quad f \in C(X), \tag{1.4.3}$$

and the following Heisenberg analogue of the Connes integration formula for Riemannian manifolds (see e.g. Chapter 3 in [46]).

Corollary 1.4.7 *Let (X, G) be a compact contact sub-Riemannian manifold. Let ν be a smooth positive density on X. For every $f \in C(X)$, we have*

$$\varphi(M_f(1 + \Delta_{G,\nu})^{-d-1}) = c_d \int_X f d\text{vol}_G.$$

The concrete expression for the sub-Riemannian volume vol_G in a chart $(\mathcal{U}_i, h_i)$ is given by the following theorem.

Theorem 1.4.8 *For every $i \in \mathbb{I}$, we have*

$$d(\mathrm{vol}_G \circ h_i^{-1})(p) = 2^{1-d} \beth(g_i^{-1}(p)) \cdot \sqrt{\det(g_i(p))} dp. \tag{1.4.4}$$

Here, the mapping $\beth : \mathrm{GL}^+(2d, \mathbb{R}) \to \mathbb{R}_+$ is defined by the setting

$$\beth(A) = \int_0^\infty \det^{\frac{1}{2}}\left(\frac{\iota s \Omega A}{\sinh(\iota s \Omega A)}\right) ds,$$

where Ω is given by (1.3.5).

The sub-Riemannian volume depends on the sub-Riemannian metric G, but it does not depend on the initial choice of the density ν. Note that the sub-Riemannian volume is *not* a Popp measure!

By [4, Proposition 19], the Popp measure $\mathcal{P}_G$ in a chart $(\mathcal{U}_i, h_i)$ is given by

$$d(\mathcal{P}_G \circ h_i^{-1})(p) = \sqrt{\frac{\det(g_i(p))}{\mathrm{tr}((\iota \Omega g_i^{-1}(p))^2)}} dp.$$

So the sub-Riemannian volume vol_G coincides (up to constant) with the Popp measure $\mathcal{P}_G$ when the matrix $g_i(p)$ is independent of p. Indeed, both the sub-Riemannian volume vol_G and the Popp measure $\mathcal{P}_G$ coincide (up to constant) with the Lebesgue measure dp. In particular, this holds for the Heisenberg group equipped with the standard sub-Riemannian metric.

There exists a deep relation (see e.g. [67]) between the Connes integration formulas and Weyl type eigenvalue asymptotics, starting with papers by Birman-Solomyak. Indeed, the most general Birman-Solomyak results (see Theorem 2 in [9]) concern pseudodifferential operators with anisotropic homogeneous symbols, which make them rather similar to the Heisenberg calculus. The authors intend to explore this relation in their future works.

Chapter 2
Principal Symbol on the Heisenberg Group

2.1 Preliminaries on Heisenberg Group

2.1.1 Schatten Class, Ideals and Traces

We recall the definitions of the Schatten classes $\mathcal{L}_p(H)$ and $\mathcal{L}_{p,\infty}(H)$. Let H be any Hilbert space and $B(H)$ be the set of all bounded operators on H. Note that if A is any compact operator on H, then $|A|$ is compact, positive and therefore diagonalizable. We define the singular values $\{\mu(k;A)\}_{k=0}^{\infty}$ to be the sequence of eigenvalues of $|A|$ (counted according to multiplicity). Equivalently, $\mu(k,A)$ can be characterized by

$$\mu(k,A) = \inf\{\|A - F\|_\infty : \operatorname{rank}(F) \leq k\}.$$

For $0 < p < \infty$, a compact operator A on H is said to belong to the Schatten class $\mathcal{L}_p(H)$ if $\{\mu(k,A)\}_{k=0}^{\infty}$ is p-summable, i.e. in the sequence space l_p. Clearly, $\mathcal{L}_p(H)$ is an ideal in $B(H)$. If $p \geq 1$, then the $\mathcal{L}_p(H)$ norm is defined as

$$\|A\|_{\mathcal{L}_p(H)} := \left(\sum_{k=0}^{\infty} \mu(k,A)^p\right)^{1/p}.$$

With this norm $\mathcal{L}_p(H)$ is a Banach ideal.

For $0 < p < \infty$, the weak Schatten class $\mathcal{L}_{p,\infty}(H)$ consists of operators A on H such that $\{\mu(k,A)\}_{k=0}^{\infty}$ is in $l_{p,\infty}$. As with $\mathcal{L}_p(H)$, $\mathcal{L}_{p,\infty}(H)$ is an ideal of $\mathcal{B}(H)$. We equip $\mathcal{L}_{p,\infty}(H)$ with quasi-norm:

$$\|A\|_{\mathcal{L}_{p,\infty}(H)} = \sup_{k \geq 0}(k+1)^{1/p} \mu(k,A).$$

For convenience, we use the abbreviations $\mathcal{L}_p$ and $\mathcal{L}_{p,\infty}$ to denote $\mathcal{L}_p(H)$ and $\mathcal{L}_{p,\infty}(H)$ respectively, whenever the Hilbert space H is clear from context. More details about Schatten class can be found in e.g. [45, 46].

In this article, the ideal $\mathcal{L}_{1,\infty}$ and the traces on this ideal play crucial roles in the proof. Recall that a trace on $\mathcal{L}_{1,\infty}$ is a linear functional $\varphi : \mathcal{L}_{1,\infty} \to \mathbb{C}$ such that for any bounded operator A and for any $B \in \mathcal{L}_{1,\infty}$, we have $\varphi(AB) = \varphi(BA)$. The trace φ is called continuous when it is continuous with respect to the $\mathcal{L}_{1,\infty}$ quasinorm. Given an orthonormal basis $\{e_n\}_{n=0}^{\infty}$ on H, we define the operator $A :=$ diag $\left\{\frac{1}{n+1}\right\}_{n=0}^{\infty}$ by $\langle e_n, Ae_m \rangle = \delta_{n,m} \frac{1}{n+1}$.

We say that φ is normalized if

$$\varphi\left(\operatorname{diag}\left\{\frac{1}{n+1}\right\}_{n=0}^{\infty}\right) = 1.$$

Note that for every unitary operator U and for every $B \in \mathcal{L}_{1,\infty}$ we have $\varphi(UBU^{-1}) = \varphi(B)$, which means that the property that φ is normalized is independent of the choice of orthonormal basis.

2.1.2 Non-commutative L_p Spaces and $L_{p,\infty}$ Spaces

For the convenience of those readers who are not familiar with non-commutative analysis, we recall definitions of the classical non-commutative L_p spaces and the weak L_p spaces, which generalise the notion of Schatten classes $\mathcal{L}_p$ and $\mathcal{L}_{p,\infty}$ defined in the preceding subsection. We refer the reader to [28] for detailed exposition.

Let $\mathcal{M} \subset B(H)$ be a von Neumann algebra equipped with a faithful normal semifinite trace τ. A closed and densely defined operator $A : \operatorname{dom}(A) \to H$ is said to be affiliated with $\mathcal{M}$ if it commutes with the commutant of $\mathcal{M}$. Besides, we say that A is τ-measurable if it is affiliated with $\mathcal{M}$ and for every $\varepsilon > 0$, there exists a projection $p \in \mathcal{M}$ such that $(1-p)H$ is contained in the domain $\operatorname{dom}(A)$ of A and $\tau(p) < \varepsilon$ [28, Definition 1.2]. For simplicity, we denote the set of all τ-measurable operators by $\mathcal{S}(\mathcal{M}, \tau)$.

Let $\mu(\cdot, X)$ denote the generalised singular value function [28, Definition 2.1] of $X \in \mathcal{S}(\mathcal{M}, \tau)$. By [28, Proposition 2.2],

$$\mu(t, X) = \inf\left\{s \geq 0 : \tau(\chi_{(s,\infty)}(|X|)) \leq t\right\}.$$

For $1 \leq p < \infty$, the non-commutative L_p space associated with $(\mathcal{M}, \tau)$ is defined by

$$L_p(\mathcal{M}, \tau) := \{X \text{ affiliated with } \mathcal{M} : \tau(|X|^p)^{1/p} < \infty\}.$$

2.1 Preliminaries on Heisenberg Group

Besides, for $0 < p < \infty$, the weak non-commutative L_p spaces associated with $(\mathcal{M}, \tau)$ are defined by

$$L_{p,\infty}(\mathcal{M}, \tau) := \{X \text{ affiliated with } \mathcal{M} : \sup_{t \geq 0} t^{1/p} \mu(t, X) < +\infty\}.$$

If $0 < p < \infty$, then the space $L_{p,\infty}(\mathcal{M}, \tau)$ equipped with the quasi-norm

$$\|X\|_{L_{p,\infty}(\mathcal{M},\tau)} := \sup_{t \geq 0} t^{1/p} \mu(t, X)$$

becomes a quasi-Banach space.

2.1.3 Group von Neumann Algebra

In this subsection, we recall the definition of the group von Neumann algebra from [52, Section 7.2.1], [60, Chapter VII, Section 3]. Let G be a unimodular Lie group and let $\lambda : G \to B(L_2(G))$ be its left regular representation. That is, for any $\gamma \in G$, $\lambda(\gamma)$ is the unitary operator on $L_2(G)$ defined by

$$(\lambda(\gamma)\xi)(h) = \xi(\gamma^{-1}h), \quad h \in G, \ \xi \in L_2(G).$$

Given $f \in C_c(G)$, write $\lambda(f)$ for the operator on $L_2(G)$ defined by

$$(\lambda(f)\xi)(h) = \int_G f(h\gamma^{-1})\xi(\gamma)\,d\gamma, \quad \xi \in L_2(G), \ h \in G.$$

Observe that $\lambda(f)\lambda(g) = \lambda(f * g)$, where $f * g \in C_c(G)$ is the convolution

$$(f * g)(h) := \int_G f(h\gamma^{-1})g(\gamma)\,d\gamma, \quad f, g \in C_c(G), \ h \in G.$$

Formally, $\lambda(f)\varphi = f * \varphi$ so that $\lambda(f)$ is rightfully called a "convolution operator".

Definition 2.1.1 The group von Neumann algebra $\text{VN}(G)$ is defined to be the closure of $\lambda(C_c(G))$ in the weak operator topology of $B(L_2(G))$. Equivalently, $\text{VN}(G)$ is the weak operator topology closure of the linear span of the family $\{\lambda(\gamma)\}_{\gamma \in G}$. That is,

$$\text{VN}(G) = \lambda(G)'' \subseteq B(L_2(G)),$$

where $''$ denotes the double commutant.

Definition 2.1.2 On the algebra $\lambda(C_c(G))$, we define a functional

$$\tau(\lambda(f)) = f(0), \quad f \in C_c(G).$$

It is not obvious that τ is well-defined, indeed in principle it is possible that λ is not injective. The fact that λ is injective follows from [60, Chapter VII, Theorem 3.4], and hence τ is well-defined on the algebra $\lambda(C_c(G))$.

Note that the unimodularity of G implies that τ is a trace such that

$$\tau(\lambda(f)\lambda(g)) = \tau(\lambda(g)\lambda(f)), \quad f, g \in C_c(G).$$

Lemma 2.1.3 ([52, Proposition 7.2.8]) *The functional τ uniquely extends to a semifinite normal trace on* $\mathrm{VN}(G)$.

Now we recall from [31, Theorem B.2.32] that if G is a unimodular group of type I, then

$$L_2(G) = \int_{\widehat{G}}^{\oplus} H_s \, d\mu(s),$$

where $\widehat{G}$ is the space of equivalence classes of irreducible unitary representations (ρ_s, H_s) of G, equipped with a Plancherel measure μ. Corresponding to this is the direct integral decomposition

$$\mathrm{VN}(G) = \int_{\widehat{G}}^{\oplus} \rho_s(\mathrm{VN}(G)) \, d\mu(s).$$

That is, given $x \in \mathrm{VN}(G)$ and $\xi \in L_2(G)$ we have

$$x\xi = \int_{\widehat{G}}^{\oplus} \rho_s(x)\xi_s \, d\mu(s),$$

where ξ_s is the component of ξ in the representation $\xi = \int_{\widehat{G}}^{\oplus} \xi_s \, d\mu(s)$. For more information about direct integral decomposition, we refer the readers to [31, Appendix B] and the reference therein.

2.1.4 Schrödinger Representation of $\mathbb{H}^d$

To adapt the above direct integral decomposition to the Heisenberg group $\mathbb{H}^d$ and its group von Neumann algebra, we first recall the Schrödinger representations (see for example [31]), which are infinite dimensional unitary representations of $\mathbb{H}^d$. For $s \in \mathbb{R}^* = \mathbb{R}\backslash\{0\}$, the Schrödinger representation ρ_s of $\mathbb{H}^d$ in $L_2(\mathbb{R}^d)$ is defined by

$$\rho_s(g)\varphi(u) = e^{is(t+\frac{1}{2}x \cdot y)} e^{i\mathrm{sgn}(s)\sqrt{|s|}y \cdot u} \varphi\left(u + \sqrt{|s|}x\right), \tag{2.1.1}$$

2.1 Preliminaries on Heisenberg Group

where $\cdot$ is the usual inner product in $\mathbb{R}^d$, $g = (z, t)$, $t \in \mathbb{R}$, $z = x + \iota y$, $x, y, u \in \mathbb{R}^d$ and $\varphi \in L_2(\mathbb{R}^d)$ (See [31, Formula (6.8)]).

The unitary dual of $\mathbb{H}^d$ is summarised in [31, Equation (6.29)],

$$\widehat{\mathbb{H}^d} = \mathbb{R} \setminus \{0\}, \quad H_s := L_2(\mathbb{R}^d)$$

up to a set of zero Plancherel measure. The Plancherel measure $d\mu(s)$ is

$$d\mu(s) = c_d |s|^d \, ds,$$

where $c_d > 0$ is some constant. The representation corresponding to $s \in \widehat{\mathbb{H}^d}$ is exactly the Schrödinger representation ρ_s given by (2.1.1). It satisfies the relations [31, Section 6.3.3, p.453]

$$\rho_s(X_j) = \iota |s|^{\frac{1}{2}} p_j, \quad \rho_s(Y_j) = \iota \operatorname{sgn}(s) |s|^{\frac{1}{2}} q_j, \quad \rho_s(T) = \iota s, \quad 1 \leq j \leq d.$$

Here $p_j := -\iota \partial_{x_j}$ and $q_j := M_{x_j}$ denote the momentum and position operators on $L_2(\mathbb{R}^d)$, respectively, and we identify ρ_s with the corresponding representation of the Lie algebra of $\mathbb{H}^d$ on $L_2(\mathbb{R}^d)$. Of course, the operators X_j and Y_j, $1 \leq j \leq d$, and T are unbounded. So, one should understand the preceding display as

$$\rho_s(e^{tX_j}) = e^{\iota t |s|^{\frac{1}{2}} p_j}, \quad \rho_s(e^{tY_j}) = e^{\iota t \operatorname{sgn}(s) |s|^{\frac{1}{2}} q_j},$$

$$\rho_s(e^{tT}) = e^{\iota s t}, \quad t \in \mathbb{R}, \quad 1 \leq j \leq d.$$

The representation ρ_s can be extended to a representation of $\operatorname{VN}(\mathbb{H}^d)$ on $L_2(\mathbb{R}^d)$, and by the Stone-von Neumann theorem

$$\rho_s(\operatorname{VN}(\mathbb{H}^d)) = B(H_s) = B(L_2(\mathbb{R}^d)), \quad s \in \mathbb{R} \setminus \{0\}.$$

Since every $B(H_s)$ is the same, it follows that

$$\operatorname{VN}(\mathbb{H}^d) \cong B(L_2(\mathbb{R}^d)) \overline{\otimes} L_\infty(\mathbb{R}).$$

An explicit $*$-isomorphism is provided by the following proposition (see [29, Proposition 2.5]).

Proposition 2.1.4 *There exists a unique isomorphism of von Neumann algebras*

$$\pi : B(L_2(\mathbb{R}^d)) \overline{\otimes} L_\infty(\mathbb{R}) \to \operatorname{VN}(\mathbb{H}^d)$$

such that

$$\exp(\iota t p_j \otimes |s|^{\frac{1}{2}}) \mapsto \lambda(\exp(tX_j)), \quad \exp(\iota t q_j \otimes \operatorname{sgn}(s)|s|^{\frac{1}{2}}) \mapsto \lambda(\exp(tY_j)).$$

Here, we use the notation s for the coordinate function $s \mapsto s$. The isomorphism π is trace-preserving in the sense that if $x \in L_1(B(L_2(\mathbb{R}^d)) \overline{\otimes} L_\infty(\mathbb{R}, |s|^d ds))$, then we have $\pi(x) \in L_1(\mathrm{VN}(\mathbb{H}^d))$ and

$$\tau(\pi(x)) = c_d \int_{-\infty}^{\infty} \mathrm{Tr}(x(s)) |s|^d ds.$$

Here, x is regarded as an integrable function from $(\mathbb{R}, |s|^d ds)$ to $\mathcal{L}_1(L_2(\mathbb{R}^d))$.

Corollary 2.1.5 *We have*

$$\pi(H_d \otimes |s|) = \Delta, \quad \pi(\iota p_j H_d^{-\frac{1}{2}} \otimes 1) = R_j,$$

$$\pi(\iota q_j H_d^{-\frac{1}{2}} \otimes \mathrm{sgn}(s)) = R_{j+d}, \quad 1 \leq j \leq d.$$

Here, H_d is the d-dimensional harmonic oscillator defined as an operator on $L_2(\mathbb{R}^d)$ by the usual formula

$$H_d = \sum_{j=1}^{d} p_j^2 + q_j^2.$$

Since $H_d \otimes |s|$ is an unbounded operator affiliated with the von Neumann algebra $B(L_2(\mathbb{R}^d)) \overline{\otimes} L_\infty(\mathbb{R})$ and not contained in the domain of π, the first identity above should be interpreted as

$$\pi(e^{\iota t H_d \otimes |s|}) = e^{\iota t \Delta}, \quad t \in \mathbb{R}.$$

2.1.5 Homogeneous Group von Neumann Algebra

Besides the algebra $\mathrm{VN}(\mathbb{H}^d)$ and its representation as $B(L_2(\mathbb{R}^d)) \overline{\otimes} L_\infty(\mathbb{R})$ via π, we are also concerned with the subalgebra of all *dilation invariant* elements of $\mathrm{VN}(\mathbb{H}^d)$, denoted by $\mathrm{VN}_{\mathrm{hom}}(\mathbb{H}^d)$ and called the homogeneous group von Neumann algebra. Here, an operator $A \in \mathrm{VN}(\mathbb{H}^d)$ is dilation invariant if for all $r > 0$ we have

$$\sigma_{r^{-1}} \circ A \circ \sigma_r = A, \tag{2.1.2}$$

where $r \mapsto \sigma_r$ is the action of $\mathbb{R}_+$ induced by the anisotropic dilation $r \mapsto \delta_r$,

$$\sigma_r u(\gamma) = u(\delta_r \gamma), \quad u \in L_2(\mathbb{H}^d), \ \gamma \in \mathbb{H}^d. \tag{2.1.3}$$

2.1 Preliminaries on Heisenberg Group

From hereon, we understand $\mathbb{C}^2$ as being identified with the two-dimensional von Neumann algebra $\mathbb{C} \oplus \mathbb{C}$ (or, equivalently, l_∞^2). That is,

$$(z_1, z_2)(w_1, w_2) = (z_1 w_1, z_2 w_2), \quad (z_1, z_2), (w_1, w_2) \in \mathbb{C}^2$$

and

$$(z_1, z_2)^* = \overline{(z_1, z_2)} = (\overline{z_1}, \overline{z_2}), \quad (z_1, z_2) \in \mathbb{C}^2.$$

Let emb denote the embedding

$$\text{emb} : B(L_2(\mathbb{R}^d)) \overline{\otimes} \mathbb{C}^2 \hookrightarrow B(L_2(\mathbb{R}^d)) \bar{\otimes} L_\infty(\mathbb{R}),$$

where $\mathbb{C}^2$ is considered as a subalgebra of $L_\infty(\mathbb{R})$, according to the embedding

$$(z_1, z_2) \mapsto z_1 \chi_{(-\infty, 0)} + z_2 \chi_{(0, \infty)} \in L_\infty(\mathbb{R}), \quad (z_1, z_2) \in \mathbb{C}^2.$$

We define a $*$-homomorphism

$$\pi_{\text{red}} : B(L_2(\mathbb{R}^d)) \overline{\otimes} \mathbb{C}^2 \to \text{VN}(\mathbb{H}^d)$$

by setting

$$\pi_{\text{red}} := \pi \circ \text{emb}.$$

Its image is precisely $\text{VN}_{\text{hom}}(\mathbb{H}^d)$, which gives a $*$-isomorphism

$$\pi_{\text{red}} : B(L_2(\mathbb{R}^d)) \overline{\otimes} \mathbb{C}^2 \to \text{VN}_{\text{hom}}(\mathbb{H}^d).$$

Observing that $(-1, 1) \mapsto \text{sgn}(s)$, it follows from Corollary 2.1.5 that

$$\pi_{\text{red}} \left(\iota p_j H_d^{-\frac{1}{2}} \otimes (1, 1) \right) = R_j,$$

$$\pi_{\text{red}} \left(\iota q_j H_d^{-\frac{1}{2}} \otimes (-1, 1) \right) = R_{j+d}, \quad 1 \leq j \leq d. \tag{2.1.4}$$

Note the equality

$$\frac{T}{\Delta} = \Delta^{-\frac{1}{2}} T \Delta^{-\frac{1}{2}} = \Delta^{-\frac{1}{2}} [X_j, Y_j] \Delta^{-\frac{1}{2}} = -R_j^* R_{j+d} + R_{j+d}^* R_j.$$

It follows now from (2.1.4) that

$$\frac{T}{\Delta} = \pi_{\text{red}}\left(\left(-(\iota p_j H_d^{-\frac{1}{2}})^* \cdot \iota q_j H_d^{-\frac{1}{2}} + (\iota q_j H_d^{-\frac{1}{2}})^* \cdot \iota p_j H_d^{-\frac{1}{2}}\right) \otimes (-1, 1)\right)$$

$$= \pi_{\text{red}}\left(H_d^{-\frac{1}{2}}[q_j, p_j] H_d^{-\frac{1}{2}} \otimes (-1, 1)\right)$$

$$= \iota \pi_{\text{red}}\left(H_d^{-1} \otimes (-1, 1)\right).$$

Taking absolute values, we obtain

$$\pi_{\text{red}}\left(H_d^{-1} \otimes (1, 1)\right) = \frac{|T|}{\Delta}. \tag{2.1.5}$$

There is a natural semifinite trace τ_{hom} on the von Neumann algebra $\text{VN}_{\text{hom}}(\mathbb{H}^d)$ given by

$$\tau_{\text{hom}}(\pi_{\text{red}}(x_1 \otimes (1, 0) + x_2 \otimes (0, 1))) = \text{Tr}(x_1) + \text{Tr}(x_2),$$

$$x_1, x_2 \in \mathcal{L}_1(L_2(\mathbb{R}^d)). \tag{2.1.6}$$

2.1.6 Automorphisms of the Heisenberg Groups

We will need the description of $\text{Aut}(\mathbb{H}^d)$ (which easily follows from Proposition 5 in [42] or from Proposition 1.21 and Theorem 1.22 in [33]).

Theorem 2.1.6 *The group* $\text{Aut}(\mathbb{H}^d)$ *consists of linear mappings on* $\mathbb{H}^d$. *This group is generated by mappings of the form*

$$\begin{pmatrix} S & 0 \\ 0 & \lambda \end{pmatrix}, \quad S \in \text{GL}(2d, \mathbb{R}), \quad \lambda \in \mathbb{R}\backslash\{0\}, \quad S\Omega S^* = \lambda \Omega.$$

Here, $\Omega \in M_{2d}(\mathbb{R})$ is given by (1.3.5).

Note that $A \in \text{Aut}(\mathbb{H}^d)$ yields that its adjoint matrix A^* also belongs to $\text{Aut}(\mathbb{H}^d)$. Indeed, if $S\Omega S^* = \lambda \Omega$, then $S^* = (S\Omega)^{-1}\lambda\Omega = -\lambda\Omega S^{-1}\Omega$. Thus,

$$S^*\Omega S = -\lambda\Omega S^{-1}\Omega \cdot \Omega S = \lambda\Omega S^{-1} S = \lambda\Omega$$

and the claim follows now from Theorem 2.1.6.

We will often identify $A \in \text{Aut}(\mathbb{H}^d)$ with the corresponding matrix $S \in \text{GL}(2d, \mathbb{R})$ given by Theorem 2.1.6, denoting it by the same letter A.

2.1 Preliminaries on Heisenberg Group

For every $A \in \mathrm{Aut}(\mathbb{H}^d)$, define the linear mapping $V_A : L_2(\mathbb{H}^d) \to L_2(\mathbb{H}^d)$ by the formula $V_A \xi = \xi \circ A^*$. Conjugation by V_A, $A \in \mathrm{Aut}(\mathbb{H}^d)$ takes left invariant vector fields to left invariant vector fields.

Lemma 2.1.7 *If $A = (a_{ij})_{i,j=1}^{2d+1} \in \mathrm{Aut}(\mathbb{H}^d)$, then*

$$V_A^{-1} X_i V_A = \sum_{j=1}^{2d} a_{ij} X_j, \quad 1 \le i \le 2d.$$

Proof Let us consider the left invariant vector field X_i as an element of the Heisenberg Lie algebra $\mathfrak{h}^d$. Let $\exp : \mathfrak{h}^d \to \mathbb{H}^d$ be the exponential map. Then the function $\gamma_{X_i}(t) = \exp(tX_i) : \mathbb{R} \to \mathbb{H}^d$ is uniquely determined by the conditions $\gamma_{X_i}(t+s) = \gamma_{X_i}(t)\gamma_{X_i}(s)$ for any $t, s \in \mathbb{R}$ and $\gamma'_{X_i}(0) = X_i(0)$.

For $\xi \in C_c^\infty(\mathbb{H}^d)$ and $h \in \mathbb{H}^d$, we have

$$X_i \xi(h) = \frac{d}{dt} \xi(h \gamma_{X_i}(t)) \Big|_{t=0}.$$

Then

$$V_A^{-1} X_i V_A \xi(h) = \xi(A^*((A^*)^{-1}(h)\gamma_{X_i}(t)))\Big|_{t=0} = \frac{d}{dt} \xi(h A^*(\gamma_{X_i}(t)))\Big|_{t=0}.$$

The function $\gamma(t) = A^*(\gamma_{X_i}(t)) : \mathbb{R} \to \mathbb{H}^d$ satisfies the conditions $\gamma(t+s) = \gamma(t)\gamma(s)$ for any $t, s \in \mathbb{R}$ and

$$\gamma'(0) = A^*[X_i(0)] = \sum_{j=1}^{2d} a_{ij} X_j(0).$$

Therefore, we have

$$A^*(\exp(tX_i)) = \exp\left(t \sum_{j=1}^{2d} a_{ij} X_j\right).$$

It follows that

$$\frac{d}{dt}\xi(hA^*(\exp(tX_i)))\Big|_{t=0} = \frac{d}{dt}\xi\left(h\exp\left(t\sum_{j=1}^{2d} a_{ij} X_j\right)\right)\Big|_{t=0} = \sum_{j=1}^{2d} a_{ij} X_j \xi(h).$$

2.1.7 Sobolev Spaces

Throughout this paper, the notation $W^{m,2}(\mathbb{H}^d)$ stands for the Heisenberg version of the inhomogeneous Sobolev space. This space serves as a natural domain for differential operators of order m.

Definition 2.1.8 The set $W^{m,2}(\mathbb{H}^d)$ consists of all $f \in L_2(\mathbb{H}^d)$ such that $X_{k_1} \cdots X_{k_l} f \in L_2(\mathbb{H}^d)$ for every $1 \leq k_1, \cdots, k_l \leq 2d$ and for every $0 \leq l \leq m$. It becomes a Hilbert space when equipped with the norm

$$\|f\|_{W^{m,2}(\mathbb{H}^d)} = \Big(\sum_{\substack{1 \leq k_1,\cdots,k_l \leq 2d \\ 0 \leq l \leq m}} \|X_{k_1} \cdots X_{k_l} f\|^2_{L_2(\mathbb{H}^d)} \Big)^{\frac{1}{2}}.$$

Occasionally, we need the homogeneous Sobolev semi-norm given by the formula

$$\|f\|_{\dot{W}^{m,2}(\mathbb{H}^d)} = \Big(\sum_{1 \leq k_1,\cdots,k_m \leq 2d} \|X_{k_1} \cdots X_{k_m} f\|^2_{L_2(\mathbb{H}^d)} \Big)^{\frac{1}{2}}.$$

Proposition 2.1.9 *The Sobolev norm is equivalent to the norm*

$$f \to \|(1+\Delta)^{\frac{m}{2}} f\|_{L_2(\mathbb{H}^d)}, \quad f \in W^{m,2}(\mathbb{H}^d).$$

The homogeneous Sobolev semi-norm is equivalent to the semi-norm

$$f \to \|\Delta^{\frac{m}{2}} f\|_{L_2(\mathbb{H}^d)}, \quad f \in W^{m,2}(\mathbb{H}^d).$$

Proof The first assertion is Corollary 4.13 in [32]. By the first assertion, we have

$$\|X_{k_1} \cdots X_{k_m} f\|_{L_2(\mathbb{H}^d)} \leq c_{k_1,\cdots,k_m,d} \|(1+\Delta)^{\frac{m}{2}} f\|_{L_2(\mathbb{H}^d)}, \quad f \in W^{m,2}(\mathbb{H}^d).$$

Applying the latter inequality to $f \circ \delta_r$ (where δ_r is the anisotropic dilation) and noting that

$$X_{k_1} \cdots X_{k_m}(f \circ \delta_r) = r^m (X_{k_1} \cdots X_{k_m} f) \circ \delta_r,$$
$$(1+\Delta)^{\frac{m}{2}}(f \circ \delta_r) = ((1+r^2\Delta)^{\frac{m}{2}} f) \circ \delta_r,$$

we obtain

$$r^m \|(X_{k_1} \cdots X_{k_m} f) \circ \delta_r\|_{L_2(\mathbb{H}^d)} \leq c_{k_1,\cdots,k_m,d} \|((1+r^2\Delta)^{\frac{m}{2}} f) \circ \delta_r\|_{L_2(\mathbb{H}^d)}.$$

Thus,

$$r^m \|X_{k_1} \cdots X_{k_m} f\|_{L_2(\mathbb{H}^d)} \leq c_{k_1,\cdots,k_m,d} \|(1+r^2\Delta)^{\frac{m}{2}} f\|_{L_2(\mathbb{H}^d)}.$$

Dividing by r^m and passing $r \to \infty$, we derive the second assertion.

2.2 Existence of the Principal Symbol Map

This subsection is devoted to the proof of Theorem 1.3.2. As mentioned above, it is obtained as an application of Theorem 3.3 in [47]. The following lemmas verify the assumptions in [47, Theorem 3.3]. We will use notation introduced in Definition 1.3.1.

Lemma 2.2.1 *If $f \in \mathcal{A}_1$, $g \in \mathcal{A}_2$, then*

$$[\pi_1(f), \pi_2(g)] \in \mathcal{K}(L_2(\mathbb{H}^d)).$$

Proof If $f \in C_c^\infty(\mathbb{H}^d)$, then $f \in \dot{W}^{1,2d+2}(\mathbb{H}^d)$. By Theorem 1.1 in [29] (or, more precisely, by Proposition 4.5 in [29]), we have $[\pi_1(f), R_k] \in \mathcal{L}_{2d+2,\infty}(L_2(\mathbb{H}^d))$. In particular, $[\pi_1(f), R_k]$ is compact. Replacing f with $\bar{f}$ and taking adjoints, we conclude that $[\pi_1(f), R_k^*]$ is compact.

Now, let $f \in C_c^\infty(\mathbb{H}^d)$ and let g be in the $*$-algebra $\mathcal{A}_2$ generated by all R_k's. It follows from the preceding paragraph using linearity and the Leibniz rule that $[\pi_1(f), \pi_2(g)]$ is compact.

Now, let $f \in C_c^\infty(\mathbb{H}^d)$ and let $g \in \mathcal{A}_2$. Choose a sequence $\{g_n\}_{n\geq 0} \subset \mathcal{A}_2$ such that $g_n \to g$ as $n \to \infty$ in the uniform norm. It follows that

$$[\pi_1(f), \pi_2(g_n)] \to [\pi_1(f), \pi_2(g)], \quad n \to \infty,$$

in the uniform norm. By the preceding paragraph, the sequence on the left hand side consists of compact operators. Hence, the operator on the right hand side is also compact.

Now, let $f \in C_0(\mathbb{H}^d)$ and let $g \in \mathcal{A}_2$. Choose a sequence $\{f_n\}_{n\geq 0} \subset C_c^\infty(\mathbb{H}^d)$ such that $f_n \to f$ as $n \to \infty$ in the uniform norm. It follows that

$$[\pi_1(f_n), \pi_2(g)] \to [\pi_1(f), \pi_2(g)], \quad n \to \infty,$$

in the uniform norm. By the preceding paragraph, the sequence on the left hand side consists of compact operators. Hence, the operator on the right hand side is also compact.

Since $\mathcal{A}_1 = \mathbb{C} + C_0(\mathbb{H}^d)$, the assertion in general case follows immediately.

In this subsection, we denote by $B(p,r)$ the ball of radius r centered at p in $\mathbb{H}^d$ with respect to Koranyi metric.

Lemma 2.2.2 *If $g \in \mathrm{VN}_{\mathrm{hom}}(\mathbb{H}^d)$ is such that*

$$\pi_1(\chi_{B(0,1)})\pi_2(g) \in \mathcal{K}(L_2(\mathbb{H}^d)),$$

then $g = 0$.

Proof It is clear that $P_k A \to 0$ in the uniform norm for an arbitrary family $\{P_k\}_{k\geq 0} \subset B(L_2(\mathbb{H}^d))$ of pairwise orthogonal projections and for every $A \in \mathcal{K}(L_2(\mathbb{H}^d))$. Fix pairwise disjoint balls $\{B(p_k, \epsilon_k)\}$ which are subsets of $B(0, 1)$. By the above fact, we have

$$\pi_1(\chi_{B(p_k,\epsilon_k)})\pi_2(g) \to 0, \quad k \to \infty,$$

in the uniform norm.

Now, let us conjugate those operators by $\lambda_r(p_k)$, where λ_r is the right regular representation of $\mathbb{H}^d$. Thus,

$$\lambda_r(p_k)^{-1}\pi_1(\chi_{B(p_k,\epsilon_k)})\pi_2(g)\lambda_r(p_k) \to 0, \quad k \to \infty,$$

in the uniform norm. Since

$$\lambda_r(p_k)^{-1}\pi_1(\chi_{B(p_k,\epsilon_k)})\lambda_r(p_k) = \pi_1(\chi_{B(0,\epsilon_k)}), \quad \lambda_r(p_k)^{-1}\pi_2(g)\lambda_r(p_k) = \pi_2(g),$$

it follows that

$$\pi_1(\chi_{B(0,\epsilon_k)})\pi_2(g) \to 0, \quad k \to \infty,$$

in the uniform norm. Now, let us conjugate those operators by σ_{ϵ_k} (here, the dilation operator σ_r is defined by (2.1.3)). Thus,

$$\sigma_{\epsilon_k}^{-1}\pi_1(\chi_{B(0,\epsilon_k)})\pi_2(g)\sigma_{\epsilon_k} \to 0, \quad k \to \infty,$$

in the uniform norm. Since

$$\sigma_{\epsilon_k}^{-1}\pi_1(\chi_{B(0,\epsilon_k)})\sigma_{\epsilon_k} = \pi_1(\chi_{B(0,1)}), \quad \sigma_{\epsilon_k}^{-1}\pi_2(g)\sigma_{\epsilon_k} = \pi_2(g),$$

it follows that

$$\sigma_{\epsilon_k}^{-1}\pi_1(\chi_{B(0,\epsilon_k)})\pi_2(g)\sigma_{\epsilon_k} = \pi_1(\chi_{B(0,1)})\pi_2(g) \to 0, \quad k \to \infty,$$

in the uniform norm. Noting that the sequence in the latter display does not depend on k, we obtain

$$\pi_1(\chi_{B(0,1)})\pi_2(g) = 0,$$

which clearly yields $g = 0$.

2.2 Existence of the Principal Symbol Map

Lemma 2.2.3 *Let $\{f_k\}_{k=1}^n \subset \mathcal{A}_1$ and $\{g_k\}_{k=1}^n \subset \mathcal{A}_2$. If*

$$\sum_{k=1}^n \pi_1(f_k)\pi_2(g_k) \in \mathcal{K}(L_2(\mathbb{H}^d)), \qquad (2.2.1)$$

then

$$\sum_{k=1}^n f_k \otimes g_k = 0.$$

Proof Fix $p \in \mathbb{H}^d$. We have

$$\lambda_r(p)\Big(\sum_{k=1}^n \pi_1(f_k)\pi_2(g_k)\Big)\lambda_r(p)^{-1} \in \mathcal{K}(L_2(\mathbb{H}^d)).$$

Clearly,

$$\lambda_r(p)\pi_1(f_k)\lambda_r(p)^{-1} = \pi_1(\lambda_r(p)f_k), \quad \lambda_r(p)\pi_2(g_k)\lambda_r(p)^{-1} = \pi_2(g_k).$$

Thus,

$$\sum_{k=1}^n \pi_1(\lambda_r(p)f_k)\pi_2(g_k) \in \mathcal{K}(L_2(\mathbb{H}^d)).$$

For each $\epsilon > 0$, we have

$$\sigma_\epsilon^{-1}\pi_1(\chi_{B(0,\epsilon)})\Big(\sum_{k=1}^n \pi_1(\lambda_r(p)f_k)\pi_2(g_k)\Big)\sigma_\epsilon \in \mathcal{K}(L_2(\mathbb{H}^d)).$$

Taking into account that

$$\sigma_{\epsilon^{-1}}\pi_1(F)\sigma_\epsilon = \pi_1(\sigma_{\epsilon^{-1}}F), \quad \sigma_{\epsilon^{-1}}\pi_2(G)\sigma_\epsilon = \pi_2(G),$$

we rewrite the latter equality as

$$\pi_1(\chi_{B(0,1)})\Big(\sum_{k=1}^n \pi_1(\sigma_{\epsilon^{-1}}(\lambda_r(p)f_k))\pi_2(g_k)\Big) \in \mathcal{K}(L_2(\mathbb{H}^d)).$$

Since f_k is continuous, it follows that

$$\chi_{B(0,1)} \cdot \sigma_{\epsilon^{-1}}(\lambda_r(p)f_k) \to f_k(p)\chi_{B(0,1)}, \quad \epsilon \downarrow 0,$$

in the uniform norm. Hence,

$$\pi_1(\chi_{B(0,1)})\Big(\sum_{k=1}^{n}\pi_1(\sigma_{\epsilon^{-1}}(\lambda_r(p)f_k))\pi_2(g_k)\Big)$$

$$\to \sum_{k=1}^{n} f_k(p)\pi_1(\chi_{B(0,1)})\pi_2(g_k), \quad \epsilon \downarrow 0,$$

in the uniform norm. Thus,

$$\sum_{k=1}^{n} f_k(p)\pi_1(\chi_{B(0,1)})\pi_2(g_k) \in \mathcal{K}(L_2(\mathbb{H}^d)).$$

In other words, we have

$$\pi_1(\chi_{B(0,1)})\pi_2\Big(\sum_{k=1}^{n} f_k(p)g_k\Big) \in \mathcal{K}(L_2(\mathbb{H}^d)).$$

Using Lemma 2.2.2, we obtain

$$\sum_{k=1}^{n} f_k(p)g_k = 0.$$

Since $p \in \mathbb{H}^d$ is arbitrary, the assertion follows.

Lemma 2.2.4 *There exists a $*$-homomorphism* sym $: \Pi \to \mathcal{A}_1 \otimes_{\min} \mathcal{A}_2$ *such that*

$$\mathrm{sym}(\pi_1(f)) = f \otimes 1, \quad \mathrm{sym}(\pi_2(g)) = 1 \otimes g, \quad f \in \mathcal{A}_1, \quad g \in \mathcal{A}_2.$$

Proof Lemmas 2.2.1 and 2.2.3 show that the conditions in [47, Theorem 3.3] are met. The assertion follows now by applying [47, Theorem 3.3].

The following assertion is well-known (see e.g. Corollary 4.1.10 in [21]).

Lemma 2.2.5 *Let $\mathcal{A}$ be a C^*-algebra. Let $\pi : \mathcal{A} \to B(H)$ be an irreducible representation. One of the following mutually exclusive options holds:*

(i) *$\pi(\mathcal{A})$ does not contain any compact operator (except for 0);*
(ii) *$\pi(\mathcal{A})$ contains every compact operator.*

We now apply Lemma 2.2.5 to the C^*-algebra $\mathcal{A} = \Pi$ and infer that Π contains the ideal $\mathcal{K}(L_2(\mathbb{H}^d))$.

Lemma 2.2.6 *The algebra $\mathcal{K}(L_2(\mathbb{H}^d))$ is contained in Π and coincides with the kernel of the homomorphism* sym.

2.2 Existence of the Principal Symbol Map

Proof Since Π contains $\pi_1(\mathcal{A}_1)$, it follows that (here, X' denotes the commutant of the set $X \subset B(L_2(\mathbb{H}^d))$)

$$\Pi' \subset \left(\pi_1(\mathcal{A}_1)\right)' = \left(\pi_1(L_\infty(\mathbb{H}^d))\right)' = \pi_1(L_\infty(\mathbb{H}^d)).$$

If $A \in \Pi'$, then $A = M_f$ for some $f \in L_\infty(\mathbb{H}^d)$ and $[R_k, M_f] = 0$ for every $1 \leq k \leq d$. In particular, $[R_k, M_f] \in \mathcal{L}_{2d+2,\infty}(L_2(\mathbb{H}^d))$ for every $1 \leq k \leq d$. It follows from Theorem 1.1 in [29] that $f \in \dot{W}^{1,2d+2}(\mathbb{H}^d)$. Furthermore, it follows from Theorem 1.2 in [29] that $X_k f = 0$ for every $1 \leq k \leq d$. Hence, f is constant. Thus, Π' is trivial.

By Proposition II.6.1.8 in [10], representation $\mathrm{id} : \Pi \to B(L_2(\mathbb{H}^d))$ is irreducible.

We now demonstrate that Π contains a non-zero compact operator. By Theorem 1.1 in [29], for every $f \in C_c^\infty(\mathbb{R}^d)$, we have $[R_k, M_f] \in \mathcal{L}_{2d+2,\infty}(L_2(\mathbb{H}^d))$ for every $1 \leq k \leq d$. This commutator is a compact element of Π. By Theorem 1.2 in [29], we have $[R_k, M_f] \neq 0$. The first assertion of the lemma follows now from Lemma 2.2.5.

Let $q : B(L_2(\mathbb{H}^d)) \to B(L_2(\mathbb{H}^d))/\mathcal{K}(L_2(\mathbb{H}^d))$ be the canonical quotient map. Recall (see the proof of Theorem 3.3 in [47]) that sym is constructed as a composition

$$\mathrm{sym} = \theta^{-1} \circ q,$$

where θ^{-1} is some linear isomorphism (its definition and properties are irrelevant at the current proof). It follows that the kernel of sym coincides with the kernel of q, which is $\mathcal{K}(L_2(\mathbb{H}^d))$.

Proof of Theorem 1.3.2 The existence is established in Lemma 2.2.4. By Lemma 2.2.6, $\mathcal{K}(L_2(\mathbb{H}^d)) \subset \Pi$ coincides with the kernel of sym.

It remains to demonstrate the surjectivity of sym. Denote the image of sym by A. By Theorem 4.1.9 in [40], $*$-homomorphic image of a C^*-algebra is again a C^*-algebra. It follows that A is a C^*-subalgebra in $\mathcal{A}_1 \otimes_{\min} \mathcal{A}_2$. Note that any element $y \in \mathcal{A}_1 \otimes \mathcal{A}_2$ belongs to A. Indeed, given

$$y = \sum_{l=1}^{L} f_l \otimes g_l \in \mathcal{A}_1 \otimes \mathcal{A}_2,$$

we have $y = \mathrm{sym}(x)$ with

$$x = \sum_{l=1}^{L} \pi_1(f_l)\pi_2(g_l) \in \Pi.$$

Since $\mathcal{A}_1 \otimes \mathcal{A}_2$ is dense in $\mathcal{A}_1 \otimes_{\min} \mathcal{A}_2$, A coincides with $\mathcal{A}_1 \otimes_{\min} \mathcal{A}_2$ and sym is surjective.

2.3 Connes Trace Formula for the Heisenberg Sub-Laplacian

This subsection is devoted to the proof of Theorem 1.4.1. We start with two technical lemmas.

Lemma 2.3.1 *If $\phi \in C_c^\infty(\mathbb{H}^d)$, then*

$$[(1+\Delta)^{-\frac{1}{2}}, M_\phi] \in \mathcal{L}_{d+1,\infty}.$$

Proof The proof is similar to (but simpler than) the one in Section 7 in [29].

Lemma 2.3.2 *If $\phi \in C_c^\infty(\mathbb{H}^d)$, then*

$$(\Delta^{-\frac{1}{2}} - (1+\Delta)^{-\frac{1}{2}})M_\phi \in \mathcal{L}_{d+1,\infty}, \quad M_\phi \Delta^{-\frac{1}{2}} \in \mathcal{L}_{2d+2,\infty}.$$

Proof The second assertion follows from Theorem 1.1 in [48].

To see the first assertion in the case $d > 1$, we note that

$$\sup_{t>0} t \cdot (t^{-\frac{1}{2}} - (t+1)^{-\frac{1}{2}}) < \infty.$$

We now write

$$(\Delta^{-\frac{1}{2}} - (1+\Delta)^{-\frac{1}{2}})M_\phi = \Delta(\Delta^{-\frac{1}{2}} - (1+\Delta)^{-\frac{1}{2}}) \cdot \Delta^{-1} M_\phi.$$

Here, the first factor is bounded and the second factor belongs to $\mathcal{L}_{d+1,\infty}$ by Theorem 1.1 in [48].

To see the first assertion in the case $d = 1$, we consider the function

$$g : t \to t^{-\frac{1}{2}} - (t+1)^{-\frac{1}{2}}, \quad t > 0.$$

By Lemma 4.2 in [48],

$$\|g(\Delta)\|_{L_2(\mathrm{VN}(\mathbb{H}^d,\tau))} = \|g\|_{L_2(\mathbb{R}_+, t^{\frac{d_{\mathrm{hom}}-2}{2}} dt)} = \|g\|_{L_2(\mathbb{R}_+, t\,dt)} < \infty.$$

By Lemma 3.2 in [48],

$$\|g(\Delta) M_\phi\|_{\mathcal{L}_2} = \|g(\Delta)\|_{L_2(\mathrm{VN}(\mathbb{H}^d,\tau))} \|\phi\|_{L_2(\mathbb{H}^d)} < \infty.$$

Thus, $g(\Delta) M_\phi \in \mathcal{L}_2 \subset \mathcal{L}_{2,\infty}$.

2.3 Connes Trace Formula for the Heisenberg Sub-Laplacian

Lemma 2.3.3 *If $\phi \in C_c^\infty(\mathbb{H}^d)$, then*

$$M_{\phi^{2d+2}}(1+\Delta)^{-d-1} - \left(\Delta^{-\frac{1}{2}} M_{\phi^2} \Delta^{-\frac{1}{2}}\right)^{d+1} \in \mathcal{L}_{\frac{2d+2}{2d+3},\infty},$$

$$M_\phi \left(M_\phi \Delta^{-1} M_\phi\right)^d M_\phi - \left(\Delta^{-\frac{1}{2}} M_{\phi^2} \Delta^{-\frac{1}{2}}\right)^{d-1} \cdot \Delta^{-\frac{1}{2}} M_{\phi^4} \Delta^{-\frac{1}{2}} \in \mathcal{L}_{\frac{2d+2}{2d+1},\infty}.$$

Proof Denote for brevity $V = M_{\phi^2}(1+\Delta)^{-1} \in \mathcal{L}_{d+1,\infty}$.

Step 1: We have

$$\Delta^{-\frac{1}{2}} M_{\phi^2} \Delta^{-\frac{1}{2}} - V, \; M_\phi \Delta^{-1} M_\phi - V \in \mathcal{L}_{\frac{2d+2}{3},\infty}.$$

Indeed,

$$\Delta^{-\frac{1}{2}} M_{\phi^2} \Delta^{-\frac{1}{2}} - (1+\Delta)^{-\frac{1}{2}} M_{\phi^2}(1+\Delta)^{-\frac{1}{2}}$$
$$= (\Delta^{-\frac{1}{2}} - (1+\Delta)^{-\frac{1}{2}}) M_\phi \cdot M_\phi \Delta^{-\frac{1}{2}}$$
$$+ (1+\Delta)^{-\frac{1}{2}} M_\phi \cdot M_\phi (\Delta^{-\frac{1}{2}} - (1+\Delta)^{-\frac{1}{2}}).$$

By Lemma 2.3.2,

$$(\Delta^{-\frac{1}{2}} - (1+\Delta)^{-\frac{1}{2}}) M_\phi \in \mathcal{L}_{d+1,\infty}, \quad M_\phi \Delta^{-\frac{1}{2}} \in \mathcal{L}_{2d+2,\infty}.$$

This yields (taking Hölder inequality into account)

$$\Delta^{-\frac{1}{2}} M_{\phi^2} \Delta^{-\frac{1}{2}} - (1+\Delta)^{-\frac{1}{2}} M_{\phi^2}(1+\Delta)^{-\frac{1}{2}} \in \mathcal{L}_{\frac{2d+2}{3},\infty}.$$

On the other hand, we have

$$(1+\Delta)^{-\frac{1}{2}} M_{\phi^2}(1+\Delta)^{-\frac{1}{2}} - M_{\phi^2}(1+\Delta)^{-1} = [(1+\Delta)^{-\frac{1}{2}}, M_\phi] \cdot M_\phi (1+\Delta)^{-\frac{1}{2}}$$
$$+ M_\phi (1+\Delta)^{-\frac{1}{2}} \cdot [M_\phi, (1+\Delta)^{-\frac{1}{2}}]$$
$$+ M_\phi \cdot [(1+\Delta)^{-1}, M_\phi].$$

The first and second summands on the right hand side belong to $\mathcal{L}_{\frac{2d+2}{3},\infty}$ by Lemma 2.3.1. The third summand on the right hand side is written as

$$M_\phi \cdot [(1+\Delta)^{-1}, M_\phi] = -M_\phi (1+\Delta)^{-1} \cdot [\Delta, M_\phi](1+\Delta)^{-1}$$
$$= \sum_{k=1}^{2d} M_\phi (1+\Delta)^{-1} \cdot [X_k^2, M_\phi](1+\Delta)^{-1}$$

$$= \sum_{k=1}^{2d} M_\phi (1+\Delta)^{-1} X_k \cdot M_{X_k\phi}(1+\Delta)^{-1}$$

$$+ \sum_{k=1}^{2d} M_\phi(1+\Delta)^{-1} \cdot M_{X_k\phi} X_k (1+\Delta)^{-1}.$$

Hence, the third summand belongs to $\mathcal{L}_{\frac{2d+2}{3},\infty}$. Thus,

$$(1+\Delta)^{-\frac{1}{2}} M_{\phi^2}(1+\Delta)^{-\frac{1}{2}} - M_{\phi^2}(1+\Delta)^{-1} \in \mathcal{L}_{\frac{2d+2}{3},\infty}.$$

Combining this with the preceding paragraph, we obtain the first inclusion in Step 1. The second inclusion in Step 1 is elementary and its proof is omitted.

Step 2: We claim

$$M_{\phi^{2n}}(1+\Delta)^{-n} - V^n \in \mathcal{L}_{\frac{2d+2}{2n+1},\infty}, \quad n \in \mathbb{N}.$$

We prove the claim by induction on n. The base case (i.e., the case $n = 1$) is immediate. It remains to establish the induction step. Suppose the claim holds for n. We have

$$M_{\phi^{2n+2}}(1+\Delta)^{-n-1} - V^{n+1} = (M_{\phi^{2n}}(1+\Delta)^{-n} - V^n) \cdot V$$

$$+ M_{\phi^{2n}} \cdot [M_{\phi^2}, (1+\Delta)^{-n-1}]$$

$$+ M_{\phi^{2n}}(1+\Delta)^{-n} \cdot [(1+\Delta)^{-1}, M_\phi^2].$$

The first summand on the right hand side belongs to $\mathcal{L}_{\frac{2d+2}{2n+3},\infty}$ by the inductive hypothesis and by the Hölder inequality. The second and third summands on the right hand side clearly belong to $\mathcal{L}_{\frac{2d+2}{2n+3},\infty}$. Hence, the left hand side belongs to $\mathcal{L}_{\frac{2d+2}{2n+3},\infty}$. This establishes the induction step and, hence, the claim.

Step 3: By Step 1, we have

$$\left(\Delta^{-\frac{1}{2}} M_{\phi^2} \Delta^{-\frac{1}{2}}\right)^{d+1} - V^{d+1} \in \mathcal{L}_{\frac{2d+2}{2d+3},\infty}.$$

The first assertion of the lemma follows now from Step 2.

Step 4: By Step 1, we have

$$\left(\Delta^{-\frac{1}{2}} M_{\phi^2} \Delta^{-\frac{1}{2}}\right)^{d-1} - V^{d-1} \in \mathcal{L}_{\frac{2d+2}{2d-1},\infty}.$$

By Step 1, we have

$$\Delta^{-\frac{1}{2}} M_{\phi^4} \Delta^{-\frac{1}{2}} - M_{\phi^4}(1+\Delta)^{-1} \in \mathcal{L}_{\frac{2d+2}{3},\infty}.$$

2.3 Connes Trace Formula for the Heisenberg Sub-Laplacian

Thus,

$$\left(\Delta^{-\frac{1}{2}} M_{\phi^2} \Delta^{-\frac{1}{2}}\right)^{d-1} \cdot \Delta^{-\frac{1}{2}} M_{\phi^4} \Delta^{-\frac{1}{2}} - V^{d-1} M_{\phi^2} V \in \mathcal{L}_{\frac{2d+2}{2d+1},\infty}.$$

Noting that

$$V^{d-1} M_{\phi^2} V - M_{\phi} V^d M_{\phi} \in \mathcal{L}_{\frac{2d+2}{2d+1},\infty},$$

we infer that

$$\left(\Delta^{-\frac{1}{2}} M_{\phi^2} \Delta^{-\frac{1}{2}}\right)^{d-1} \cdot \Delta^{-\frac{1}{2}} M_{\phi^4} \Delta^{-\frac{1}{2}} - M_{\phi} V^d M_{\phi} \in \mathcal{L}_{\frac{2d+2}{2d+1},\infty}.$$

The second assertion of the lemma follows now from Step 1.

We will identify $C^*(\{R_k\}_{k=1}^{2d})$ with $B(L_2(\mathbb{R}^d)) \otimes \mathbb{C}^2$ by means of π_{red} (see Sect. 2.1.5). By (2.1.6), the $*$-isomorphism π_{red} identifies the trace τ_{hom} with the trace $\text{Tr} \otimes \Sigma$ on $B(L_2(\mathbb{R}^d)) \otimes \mathbb{C}^2$ given by the tensor product of the standard operator trace on $B(L_2(\mathbb{R}^d))$ and the "sum" functional

$$\Sigma : \mathbb{C}^2 \to \mathbb{C}, \quad \Sigma(\xi_1, \xi_2) = \xi_1 + \xi_2, \quad (\xi_1, \xi_2) \in \mathbb{C}^2.$$

Recall that H_d is the harmonic oscillator on $\mathbb{R}^d$.

Lemma 2.3.4 *Let $f \in L_\infty(\mathbb{H}^d)$ be compactly supported. Let $x \in B(L_2(\mathbb{R}^d)) \otimes \mathbb{C}^2$. For every continuous normalised trace φ on $\mathcal{L}_{1,\infty}$, we have*

$$\varphi(\pi_{\text{red}}(x) M_f (1+\Delta)^{-d-1}) = c_d \int_{\mathbb{H}^d} f \cdot (\text{Tr} \otimes \Sigma)(x (H_d^{-d-1} \otimes (1,1))).$$

Proof Without loss of generality, we may assume that $f \geq 0$. Choose $0 \leq \phi \in C_c^\infty(\mathbb{H}^d)$ such that $f = f\phi$. We write

$$\pi_{\text{red}}(x) M_f (1+\Delta)^{-d-1} = \pi_{\text{red}}(x) M_f \cdot M_{\phi^{2d+2}} (1+\Delta)^{-d-1}.$$

By Lemma 2.3.3, we have

$$\varphi(\pi_{\text{red}}(x) M_f (1+\Delta)^{-d-1}) = \varphi\left(\pi_{\text{red}}(x) M_f \left(\Delta^{-\frac{1}{2}} M_{\phi^2} \Delta^{-\frac{1}{2}}\right)^{d+1}\right).$$

Set

$$A = \pi_{\text{red}}(x) M_f \Delta^{-\frac{1}{2}} M_\phi \left(M_\phi \Delta^{-1} M_\phi\right)^d \in \mathcal{L}_{\frac{2d+2}{2d+1},\infty}, \quad B = M_\phi \Delta^{-\frac{1}{2}} \in \mathcal{L}_{2d+2,\infty}.$$

Note that

$$\pi_{\text{red}}(x) M_f \left(\Delta^{-\frac{1}{2}} M_{\phi^2} \Delta^{-\frac{1}{2}} \right)^{d+1} = AB.$$

Recall that (see e.g. p.3 in [22]) $[\mathcal{L}_{2d+2,\infty}, \mathcal{L}_{\frac{2d+2}{2d+1},\infty}] = [\mathcal{L}_{1,\infty}, \mathcal{L}_{\infty}]$. Hence, φ vanishes on the subspace $[\mathcal{L}_{2d+2,\infty}, \mathcal{L}_{\frac{2d+2}{2d+1},\infty}]$. By $\varphi(AB) = \varphi(BA)$, we, therefore, have

$$\varphi\left(\pi_{\text{red}}(x) M_f \left(\Delta^{-\frac{1}{2}} M_{\phi^2} \Delta^{-\frac{1}{2}} \right)^{d+1} \right)$$
$$= \varphi\left(M_\phi \Delta^{-\frac{1}{2}} \cdot \pi_{\text{red}}(x) M_f \Delta^{-\frac{1}{2}} \cdot M_\phi \left(M_\phi \Delta^{-1} M_\phi \right)^d \right)$$
$$= \varphi\left(M_\phi \cdot \Delta^{-\frac{1}{2}} \pi_{\text{red}}(x) M_f \cdot M_\phi \Delta^{-\frac{1}{2}} \cdot M_\phi \left(M_\phi \Delta^{-1} M_\phi \right)^d \right).$$

Since

$$\Delta^{-\frac{1}{2}} \pi_{\text{red}}(x) M_f \in \mathcal{L}_{2d+2,\infty},$$

it follows that

$$\Delta^{-\frac{1}{2}} \pi_{\text{red}}(x) M_f \cdot M_\phi \Delta^{-\frac{1}{2}} \cdot M_\phi \left(M_\phi \Delta^{-1} M_\phi \right)^d \in \mathcal{L}_{1,\infty}.$$

By the tracial property, we have

$$\varphi\left(M_\phi \cdot \Delta^{-\frac{1}{2}} \pi_{\text{red}}(x) M_f \cdot M_\phi \Delta^{-\frac{1}{2}} \cdot M_\phi \left(M_\phi \Delta^{-1} M_\phi \right)^d \right)$$
$$= \varphi\left(\Delta^{-\frac{1}{2}} \pi_{\text{red}}(x) M_f \cdot M_\phi \Delta^{-\frac{1}{2}} \cdot M_\phi \left(M_\phi \Delta^{-1} M_\phi \right)^d M_\phi \right).$$

Thus,

$$\varphi(\pi_{\text{red}}(x) M_f (1+\Delta)^{-d-1})$$
$$= \varphi\left(\Delta^{-\frac{1}{2}} \pi_{\text{red}}(x) M_f \cdot M_\phi \Delta^{-\frac{1}{2}} \cdot M_\phi \left(M_\phi \Delta^{-1} M_\phi \right)^d M_\phi \right).$$

By Lemma 2.3.3, we have

$$\varphi(\pi_{\text{red}}(x) M_f (1+\Delta)^{-d-1})$$
$$= \varphi\left(\Delta^{-\frac{1}{2}} \pi_{\text{red}}(x) M_f \cdot M_\phi \Delta^{-\frac{1}{2}} \cdot \left(\Delta^{-\frac{1}{2}} M_{\phi^2} \Delta^{-\frac{1}{2}} \right)^{d-1} \cdot \Delta^{-\frac{1}{2}} M_{\phi^4} \Delta^{-\frac{1}{2}} \right).$$

2.3 Connes Trace Formula for the Heisenberg Sub-Laplacian

Now, we use the notation

$$A(x, f) = M_f \pi_{\text{red}}(x) |T|^{-\frac{1}{2}}$$

used in [29]. Setting

$$f_1 = f, \quad f_k = \phi, \quad 2 \le k \le 2d, \quad f_{2d+1} = f_{2d+2} = \phi^2,$$

$$x_1 = x^*(H_d^{-\frac{1}{2}} \otimes (1,1)), \quad x_k = H_d^{-\frac{1}{2}} \otimes (1,1), \quad 2 \le k \le 2d+2,$$

we write

$$\Delta^{-\frac{1}{2}} \pi_{\text{red}}(x) M_f \cdot M_\phi \Delta^{-\frac{1}{2}} \cdot \left(\Delta^{-\frac{1}{2}} M_{\phi^2} \Delta^{-\frac{1}{2}}\right)^{d-1} \cdot \Delta^{-\frac{1}{2}} M_{\phi^4} \Delta^{-\frac{1}{2}}$$
$$= A(x_1, f_1)^* A(x_2, f_2) A(x_3, f_3)^* A(x_4, f_4)$$
$$\cdots A(x_{2d+1}, f_{2d+1})^* A(x_{2d+2}, f_{2d+2}).$$

Hence,

$$\varphi(\pi_{\text{red}}(x) M_f (1+\Delta)^{-d-1})$$
$$= \varphi(A(x_1, f_1)^* A(x_2, f_2) A(x_3, f_3)^* A(x_4, f_4)$$
$$\cdots A(x_{2d+1}, f_{2d+1})^* A(x_{2d+2}, f_{2d+2})).$$

Using Theorem 6.1 in [29], we write

$$\varphi(\pi_{\text{red}}(x) M_f (1+\Delta)^{-d-1})$$
$$= c_d \left(\int_{\mathbb{H}^d} \bar{f}_1 f_2 \bar{f}_3 f_4 \cdots \bar{f}_{2d+1} f_{2d+2}\right) \cdot (\text{Tr} \otimes \Sigma)\left(x_1^* x_2 x_3^* x_4 \cdots x_{2d+1}^* x_{2d+2}\right).$$

By the definition of f_k's and x_k's, we have

$$\bar{f}_1 f_2 \bar{f}_3 f_4 \cdots \bar{f}_{2d+1} f_{2d+2} = f,$$

$$x_1^* x_2 x_3^* x_4 \cdots x_{2d+1}^* x_{2d+2} = (H_d^{-\frac{1}{2}} \otimes (1,1)) x (H_d^{-d-\frac{1}{2}} \otimes (1,1)).$$

The assertion follows now from the tracial property.

Lemma 2.3.5 *For every compactly supported $S \in \Pi$, we have*

$$\varphi(S(1+\Delta)^{-d-1}) = c_d \left(\int_{\mathbb{H}^d} \otimes \tau_{\text{hom}}\right)\left(\text{sym}(S) \cdot \left(1 \otimes \frac{|T|}{\Delta}\right)^{d+1}\right). \quad (2.3.1)$$

Proof Let $f \in C_c(\mathbb{H}^d)$ and let $g \in C^*(\{R_k\}_{k=1}^{2d})$. Choose $x \in B(L_2(\mathbb{R}^d)) \otimes \mathbb{C}^2$ such that $g = \pi_{\mathrm{red}}(x)$. It follows from Lemma 2.3.4 that

$$\varphi(\pi_2(g)\pi_1(f)(1+\Delta)^{-d-1}) = \varphi(\pi_{\mathrm{red}}(x)M_f(1+\Delta)^{-d-1})$$

$$= c_d \int_{\mathbb{H}^d} f \cdot (\mathrm{Tr} \otimes \Sigma)(x(H_d^{-d-1} \otimes (1,1)))$$

$$= c_d \int_{\mathbb{H}^d} f \cdot \tau_{\mathrm{hom}}\left(g \cdot (\frac{|T|}{\Delta})^{d+1}\right).$$

The assertion follows from Lemma 2.3.4 and Theorem 5.2.8 in [46].

The following lemma relates the traces τ_{hom} and τ and allows us to rewrite the right hand side of (2.3.1) in terms of τ.

Lemma 2.3.6 *Let $x \in \mathrm{VN}_{\mathrm{hom}}(\mathbb{H}^d)$ and let y be affiliated with $\mathrm{VN}_{\mathrm{hom}}(\mathbb{H}^d)$. Suppose y is positive, boundedly invertible and $y^{-1} \in L_{d,\infty}(\mathrm{VN}_{\mathrm{hom}}(\mathbb{H}^d))$. We have*

$$\tau_{\mathrm{hom}}(xy^{-d-1}) = c_d \tau(xe^{-|T|y}).$$

Proof We can write

$$x = \pi_{\mathrm{red}}(x_1 \otimes (1,0) + x_2 \otimes (0,1)), \quad y = \pi_{\mathrm{red}}(y_1 \otimes (1,0) + y_2 \otimes (0,1)).$$

By (2.1.6), we have

$$\tau_{\mathrm{hom}}(xy^{-d-1}) = \mathrm{Tr}(x_1 y_1^{-d-1} + x_2 y_2^{-d-1})$$

and, by (2.6.1),

$$\tau(xe^{-|T|y}) = \int_0^\infty \mathrm{Tr}(x_1 e^{-|s|y_1} + x_2 e^{-|s|y_2})|s|^d ds.$$

The assertion follows now from the abstract equality

$$\int_{\mathbb{R}} e^{-|s|X}|s|^d ds = 2\Gamma(d+1)X^{-d-1}.$$

Proof of Theorem 1.4.1 We have

$$(\int_{\mathbb{H}^d} \otimes \tau_{\mathrm{hom}})\left(\mathrm{sym}(S) \cdot \left(1 \otimes \frac{|T|}{\Delta}\right)^{d+1}\right) = \int_{\mathbb{H}^d} \tau_{\mathrm{hom}}\left(\mathrm{sym}(S)(p) \cdot \left(\frac{|T|}{\Delta}\right)^{d+1}\right) dp.$$

2.4 Elliptic Estimate

Applying Lemma 2.3.6 with

$$x = \operatorname{sym}(S)(p), \quad y = \frac{\Delta}{|T|},$$

we obtain

$$\tau_{\operatorname{hom}}\left(\operatorname{sym}(S)(p) \cdot \left(\frac{|T|}{\Delta}\right)^{d+1}\right) = c_d \tau\left(\operatorname{sym}(S)(p) \cdot e^{-\Delta}\right), \quad p \in \mathbb{H}^d$$

and, therefore,

$$\left(\int_{\mathbb{H}^d} \otimes \tau_{\operatorname{hom}}\right)\left(\operatorname{sym}(S) \cdot \left(1 \otimes \frac{|T|}{\Delta}\right)^{d+1}\right) = c_d \int_{\mathbb{H}^d} \tau\left(\operatorname{sym}(S)(p) \cdot e^{-\Delta}\right) dp$$

$$= c_d \left(\int_{\mathbb{H}^d} \otimes \tau\right)\left(\operatorname{sym}(S) \cdot (1 \otimes e^{-\Delta})\right).$$

The assertion follows now from Lemma 2.3.5.

2.4 Elliptic Estimate

Our next goal is to extend the Connes trace formula for the Heisenberg sub-Laplacian, Theorem 1.4.1, to an arbitrary elliptic, self-adjoint and positive differential operator of order 2 (in Heisenberg sense) on the Heisenberg group. This will be done in Sect. 2.6 (see Theorem 2.6.1). Before, in this section and Sect. 2.5, we will establish some general results on elliptic operators on the Heisenberg group.

Definition 2.4.1 We say that P is a differential operator of order m (in Heisenberg sense) on $\mathbb{H}^d$ if

$$P = \sum_{\substack{1 \leq k_1, \cdots, k_l \leq 2d \\ 0 \leq l \leq m}} M_{a_{k_1, \cdots, k_l}} X_{k_1} \cdots X_{k_l},$$

where the coefficients $a_{k_1, \cdots, k_l}$ are smooth functions on $\mathbb{H}^d$.

Note that such a representation is not necessarily unique.

Definition 2.4.2 Let P be a differential operator of order m on $\mathbb{H}^d$. We write $P = Q + R$, where R is a differential operator of order $m - 1$ and

$$Q = \sum_{1 \leq k_1, \cdots, k_m \leq 2d} M_{a_{k_1, \cdots, k_m}} X_{k_1} \cdots X_{k_m}.$$

For every $p \in \mathbb{H}^d$, define a left-invariant differential operator Q_p on $\mathbb{H}^d$ by the formula

$$Q_p = \sum_{1 \leq k_1, \cdots, k_m \leq 2d} a_{k_1, \cdots, k_m}(p) X_{k_1} \cdots X_{k_m}.$$

We say that P is uniformly elliptic if there exists $c_P^{(1)} > 0$ such that

$$\|Q_p f\|_{L_2(\mathbb{H}^d)} \geq c_P^{(1)} \|f\|_{\dot{W}^{m,2}(\mathbb{H}^d)}, \quad f \in W^{m,2}(\mathbb{H}^d), \quad p \in \mathbb{H}^d.$$

The following theorem is an analogue of the standard elliptic estimate.

Theorem 2.4.3 *Let P be a uniformly elliptic differential operator of order m on $\mathbb{H}^d$. Suppose the coefficients of P are smooth and constant outside some ball. We have*

$$\|Pf\|_{L_2(\mathbb{H}^d)} + \|f\|_{L_2(\mathbb{H}^d)} \geq c_P \|f\|_{W^{m,2}(\mathbb{H}^d)}, \quad f \in W^{m,2}(\mathbb{H}^d).$$

The rest of this subsection is devoted to the proof of Theorem 2.4.3.
The next lemma is the traditional Schauder-type argument.

Lemma 2.4.4 *Let P be as in Theorem 2.4.3. There exists $\epsilon > 0$ such that*

$$\|Pf\|_{L_2(\mathbb{H}^d)} + \|f\|_{L_2(\mathbb{H}^d)} \geq c_P \|f\|_{W^{m,2}(L_2(\mathbb{H}^d))}, \quad f \in W^{m,2}(\mathbb{H}^d),$$

provided that $\mathrm{diam}(\mathrm{supp}(f)) < \epsilon$. *Here the diameter is with respect to the Carnot-Caratheodory distance (see Example 1.3.3).*

Proof By the ellipticity P, there exists $c_P^{(1)} > 0$ such that

$$\|Q_p h\|_{L_2(\mathbb{H}^d)} \geq c_P^{(1)} \|h\|_{\dot{W}^{m,2}(\mathbb{H}^d)}, \quad h \in W^{m,2}(\mathbb{H}^d), \quad p \in \mathbb{H}^d.$$

Choose $\epsilon > 0$ such that

$$\sum_{1 \leq k_1, \cdots, k_m \leq 2d} |a_{k_1, \cdots, k_m}(p_1) - a_{k_1, \cdots, k_m}(p_2)| < \frac{1}{2} c_P^{(1)}$$

whenever $\mathrm{dist}(p_1, p_2) < \epsilon$. Here dist denotes the Carnot-Caratheodory distance.

Let $f \in W^{m,2}(\mathbb{H}^d)$ be such that $\mathrm{diam}(\mathrm{supp}(f)) < \epsilon$. Choose $p \in \mathrm{supp}(f)$. We have

$$\|Qf - Q_p f\|_{L_2(\mathbb{H}^d)}$$
$$\leq \sum_{1 \leq k_1, \cdots, k_m \leq 2d} \|M_{a_{k_1, \cdots, k_m} - a_{k_1, \cdots, k_m}(p)} X_{k_1} \cdots X_{k_m} f\|_{L_2(\mathbb{H}^d)}$$

2.4 Elliptic Estimate

$$\leq \sum_{1 \leq k_1, \cdots, k_m \leq 2d} \|a_{k_1, \cdots, k_m} - a_{k_1, \cdots, k_m}(p)\|_{L_\infty(\mathrm{supp}(f))} \|X_{k_1} \cdots X_{k_m} f\|_{L_2(\mathbb{H}^d)}$$

$$\leq \sum_{1 \leq k_1, \cdots, k_m \leq 2d} \|a_{k_1, \cdots, k_m} - a_{k_1, \cdots, k_m}(p)\|_{L_\infty(\mathrm{supp}(f))} \cdot \|f\|_{\dot{W}^{m,2}(\mathbb{H}^d)}$$

$$\leq \frac{1}{2} c_P^{(1)} \|f\|_{\dot{W}^{m,2}(\mathbb{H}^d)}.$$

Consequently,

$$\|Qf\|_{L_2(\mathbb{H}^d)} \geq \|Q_p f\|_{L_2(\mathbb{H}^d)} - \|(Q - Q_p)f\|_{L_2(\mathbb{H}^d)} \geq \frac{1}{2} c_P^{(1)} \|f\|_{\dot{W}^{m,2}(\mathbb{H}^d)}.$$

Thus,

$$\|Pf\|_{L_2(\mathbb{H}^d)} \geq \|Qf\|_{L_2(\mathbb{H}^d)} - \|Rf\|_{L_2(\mathbb{H}^d)}$$

$$\geq \frac{1}{2} c_P^{(1)} \|f\|_{\dot{W}^{m,2}(\mathbb{H}^d)} - c_P^{(2)} \|f\|_{W^{m-1,2}(\mathbb{H}^d)}$$

$$\geq \frac{1}{2} c_P^{(1)} \|f\|_{W^{m,2}(\mathbb{H}^d)} - c_P^{(3)} \|f\|_{W^{m-1,2}(\mathbb{H}^d)}.$$

By complex interpolation, we have

$$\|f\|_{W^{m-1,2}(\mathbb{H}^d)} \leq c_m \|f\|_{W^{m,2}(\mathbb{H}^d)}^{1-\frac{1}{m}} \|f\|_{L_2(\mathbb{H}^d)}^{\frac{1}{m}}.$$

Thus,

$$\|Pf\|_{L_2(\mathbb{H}^d)} \geq \frac{1}{2} c_P^{(1)} \|f\|_{W^{m,2}(\mathbb{H}^d)} - c_m c_P^{(3)} \|f\|_{W^{m,2}(\mathbb{H}^d)}^{1-\frac{1}{m}} \|f\|_{L_2(\mathbb{H}^d)}^{\frac{1}{m}}.$$

For every $\varepsilon > 0$, we have

$$\|f\|_{W^{m,2}(\mathbb{H}^d)}^{1-\frac{1}{m}} \|f\|_{L_2(\mathbb{H}^d)}^{\frac{1}{m}} \leq \varepsilon \|f\|_{W^{m,2}(\mathbb{H}^d)} + \varepsilon^{-\frac{1}{m-1}} \|f\|_{L_2(\mathbb{H}^d)}.$$

Setting $\varepsilon = \frac{1}{4} c_P^{(1)}$, we arrive at

$$\|Pf\|_{L_2(\mathbb{H}^d)} \geq \frac{1}{4} c_P^{(1)} \|f\|_{W^{m,2}(\mathbb{H}^d)} - c_{m,P} \|f\|_{L_2(\mathbb{H}^d)}.$$

This immediately yields the assertion.

For every word $w = k_1 \cdots k_{\mathrm{length}(w)}$ in the alphabet $\{1, \cdots, 2d\}$, we set

$$X_w = X_{k_1} \cdots X_{k_{\mathrm{length}(w)}}.$$

Lemma 2.4.5 *We have*

$$X_w(f_1 f_2) = \sum_{\mathscr{A} \subset \{1,\cdots,\text{length}(w)\}} X_{w_{\mathscr{A}}} f_1 \cdot X_{w_{\mathscr{A}^c}} f_2.$$

Here,

$$w_{\mathscr{A}} = k_{l_1} \cdots k_{l_s}, \quad \mathscr{A} = \{l_1, \cdots, l_s\}, \quad l_1 < l_2 < \cdots < l_s.$$

Proof We prove the assertion by induction on length(w). Suppose the assertion is proved for every word of length m. Let $w = k_1 \cdots k_{m+1}$ be a fixed word of length $m+1$. Denote $w' = k_1 \cdots k_m$. Set $f_3 = X_{k_{m+1}} f_1$ and $f_4 = X_{k_{m+1}} f_2$. We have

$$X_w(f_1 f_2) = X_{w'}(X_{k_{m+1}}(f_1 f_2))$$
$$= X_{w'}(f_1 f_4 + f_3 f_2)$$
$$= \sum_{\mathscr{A} \subset \{1,\cdots,m\}} X_{w'_{\mathscr{A}}} f_1 \cdot X_{w'_{\mathscr{A}^c}} f_4 + \sum_{\mathscr{A} \subset \{1,\cdots,m\}} X_{w'_{\mathscr{A}}} f_3 \cdot X_{w'_{\mathscr{A}^c}} f_2.$$

It is immediate that

$$X_{w'_{\mathscr{A}}} f_1 = X_{w_{\mathscr{A}}} f_1, \quad X_{w'_{\mathscr{A}^c}} f_4 = X_{w_{\mathscr{A}^c}} f_2,$$

$$X_{w'_{\mathscr{A}}} f_3 = X_{w_{\mathscr{A} \cup \{m+1\}}} f_1, \quad X_{w'_{\mathscr{A}^c}} f_2 = X_{w_{(\mathscr{A} \cup \{m+1\})^c}} f_2,$$

where the complements on the right hand side are taken in the set $\{1, \cdots, m+1\}$. Thus, (again, the complements on the right hand side are taken in the set $\{1, \cdots, m+1\}$)

$$X_w(f_1 f_2) = \sum_{\mathscr{A} \subset \{1,\cdots,m\}} X_{w_{\mathscr{A}}} f_1 \cdot X_{w_{\mathscr{A}^c}} f_2$$
$$+ \sum_{\mathscr{A} \subset \{1,\cdots,m\}} X_{w_{\mathscr{A} \cup \{m+1\}}} f_1 \cdot X_{w_{(\mathscr{A} \cup \{m+1\})^c}} f_2$$
$$= \sum_{\substack{\mathscr{A} \subset \{1,\cdots,m+1\} \\ m+1 \notin \mathscr{A}}} X_{w_{\mathscr{A}}} f_1 \cdot X_{w_{\mathscr{A}^c}} f_2 + \sum_{\substack{\mathscr{A} \subset \{1,\cdots,m+1\} \\ m+1 \in \mathscr{A}}} X_{w_{\mathscr{A}}} f_1 \cdot X_{w_{\mathscr{A}^c}} f_2$$
$$= \sum_{\mathscr{A} \subset \{1,\cdots,m+1\}} X_{w_{\mathscr{A}}} f_1 \cdot X_{w_{\mathscr{A}^c}} f_2.$$

Recall that $\mathbb{H}^d(\mathbb{Z})$ is the subgroup in $\mathbb{H}^d$ which coincides as a set with $\mathbb{Z}^{2d+1}$.

2.4 Elliptic Estimate

Lemma 2.4.6 *Fix* $0 \leq \psi \in C_c^\infty(\mathbb{H}^d)$ *be such that*

$$\sum_{n \in \mathbb{H}^d(\mathbb{Z})} \psi_n = 1, \quad \psi_k(p) = \psi(pk^{-1}), \quad p \in \mathbb{H}^d, \quad k \in \mathbb{H}^d(\mathbb{Z}).$$

Here, pk^{-1} is understood in the sense of $\mathbb{H}^d$. We have

$$\left(\sum_{n \in \mathbb{H}^d(\mathbb{Z})} \|f \psi_n\|_{W^{m,2}(\mathbb{H}^d)}^2 \right)^{\frac{1}{2}} \geq c_{m,\psi} \|f\|_{W^{m,2}(\mathbb{H}^d)}.$$

Proof Observe that $\sum_{n \in \mathbb{H}^d(\mathbb{Z})} (X_w \psi_n)^2$ is bounded for every word w in the alphabet $\{1, \cdots, 2d\}$. Also observe that

$$\sum_{n \in \mathbb{H}^d(\mathbb{Z})} \psi_n^2 \geq c_\psi > 0.$$

Let w be a word of length m in the alphabet $\{1, \cdots, 2d\}$. By Lemma 2.4.5, we have

$$X_w f \cdot \psi_n = X_w(f \psi_n) - \sum_{\mathscr{A} \subsetneq \{1,\cdots,m\}} X_{w_{\mathscr{A}}} f \cdot X_{w_{\mathscr{A}^c}} \psi_n.$$

By triangle inequality, we have

$$\|X_w f \cdot \psi_n\|_{L_2(\mathbb{H}^d)} \leq \|X_w(f \psi_n)\|_{L_2(\mathbb{H}^d)} + \sum_{\mathscr{A} \subsetneq \{1,\cdots,m\}} \|X_{w_{\mathscr{A}}} f \cdot X_{w_{\mathscr{A}^c}} \psi_n\|_{L_2(\mathbb{H}^d)}$$

$$\leq \|f \psi_n\|_{W^{m,2}(\mathbb{H}^d)} + \sum_{\mathscr{A} \subsetneq \{1,\cdots,m\}} \|X_{w_{\mathscr{A}}} f \cdot X_{w_{\mathscr{A}^c}} \psi_n\|_{L_2(\mathbb{H}^d)}.$$

Hence,

$$\|X_w f \cdot \psi_n\|_{L_2(\mathbb{H}^d)}^2 \leq 2^m \Bigg(\|f \psi_n\|_{W^{m,2}(\mathbb{H}^d)}^2 + \sum_{\mathscr{A} \subsetneq \{1,\cdots,m\}} \|X_{w_{\mathscr{A}}} f \cdot X_{w_{\mathscr{A}^c}} \psi_n\|_{L_2(\mathbb{H}^d)}^2 \Bigg).$$

Summing over $n \in \mathbb{H}^d(\mathbb{Z})$, we obtain

$$\left\| X_w f \cdot \left(\sum_{n \in \mathbb{H}^d(\mathbb{Z})} \psi_n^2 \right)^{\frac{1}{2}} \right\|_{L_2(\mathbb{H}^d)}^2$$

$$= \sum_{n \in \mathbb{H}^d(\mathbb{Z})} \|X_w f \cdot \psi_n\|_{L_2(\mathbb{H}^d)}^2$$

$$\leq 2^m \left(\sum_{n \in \mathbb{H}^d(\mathbb{Z})} \|f\psi_n\|^2_{W^{m,2}(\mathbb{H}^d)} + \sum_{\mathscr{A} \subsetneq \{1,\cdots,m\}} \sum_{n \in \mathbb{H}^d(\mathbb{Z})} \|X_{w_\mathscr{A}} f \cdot X_{w_{\mathscr{A}^c}} \psi_n\|^2_{L_2(\mathbb{H}^d)} \right)$$

$$= 2^m \left(\sum_{n \in \mathbb{H}^d(\mathbb{Z})} \|f\psi_n\|^2_{W^{m,2}(\mathbb{H}^d)} \right.$$

$$\left. + \sum_{\mathscr{A} \subsetneq \{1,\cdots,m\}} \left\| X_{w_\mathscr{A}} f \cdot \left(\sum_{n \in \mathbb{H}^d(\mathbb{Z})} (X_{w_{\mathscr{A}^c}} \psi_n)^2 \right)^{\frac{1}{2}} \right\|^2_{L_2(\mathbb{H}^d)} \right).$$

If $\mathscr{A} \subsetneq \{1, \cdots, m\}$, then

$$\left\| X_{w_\mathscr{A}} f \cdot \left(\sum_{n \in \mathbb{H}^d(\mathbb{Z})} (X_{w_{\mathscr{A}^c}} \psi_n)^2 \right)^{\frac{1}{2}} \right\|_{L_2(\mathbb{H}^d)}$$

$$\leq \|f\|_{W^{m-1,2}(\mathbb{H}^d)} \left\| \sum_{n \in \mathbb{H}^d(\mathbb{Z})} (X_{w_{\mathscr{A}^c}} \psi_n)^2 \right\|^{\frac{1}{2}}_{L_\infty(\mathbb{H}^d)}.$$

Thus,

$$c_\psi \|X_w f\|^2_{L_2(\mathbb{H}^d)} \leq \left\| X_w f \cdot \left(\sum_{n \in \mathbb{H}^d(\mathbb{Z})} \psi_n^2 \right)^{\frac{1}{2}} \right\|^2_{L_2(\mathbb{H}^d)}$$

$$\leq 2^m \left(\sum_{n \in \mathbb{H}^d(\mathbb{Z})} \|f\psi_n\|^2_{W^{m,2}(\mathbb{H}^d)} \right.$$

$$\left. + \sum_{\mathscr{A} \subsetneq \{1,\cdots,m\}} \|f\|^2_{W^{m-1,2}(\mathbb{H}^d)} \cdot \left\| \sum_{n \in \mathbb{H}^d(\mathbb{Z})} (X_{w_{\mathscr{A}^c}} \psi_n)^2 \right\|_{L_\infty(\mathbb{H}^d)} \right).$$

Summing over w with $\text{length}(w) = m$ and taking into account that

$$\sum_{\text{length}(w)=m} \|X_w f\|^2_{L_2(\mathbb{H}^d)} = \|f\|^2_{W^{m,2}(\mathbb{H}^d)} - \|f\|^2_{W^{m-1,2}(\mathbb{H}^d)},$$

we obtain

$$c_\psi \|f\|^2_{W^{m,2}(\mathbb{H}^d)}$$

$$\leq (4d)^m \sum_{n \in \mathbb{H}^d(\mathbb{Z})} \|f\psi_n\|^2_{W^{m,2}(\mathbb{H}^d)}$$

$$+ \|f\|^2_{W^{m-1,2}(\mathbb{H}^d)} \cdot \left(c_\psi + (4d)^m \sum_{\mathscr{A} \subsetneq \{1,\cdots,m\}} \left\| \sum_{n \in \mathbb{H}^d(\mathbb{Z})} (X_{w_{\mathscr{A}^c}} \psi_n)^2 \right\|_{L_\infty(\mathbb{H}^d)} \right).$$

2.4 Elliptic Estimate

In other words,

$$c_\psi \|f\|^2_{W^{m,2}(\mathbb{H}^d)} \leq (4d)^m \sum_{n \in \mathbb{H}^d(\mathbb{Z})} \|f\psi_n\|^2_{W^{m,2}(\mathbb{H}^d)} + c^{(1)}_{m,\psi} \|f\|^2_{W^{m-1,2}(\mathbb{H}^d)}.$$

By complex interpolation, we have

$$\|f\|_{W^{m-1,2}(\mathbb{H}^d)} \leq c_m \|f\|^{1-\frac{1}{m}}_{W^{m,2}(\mathbb{H}^d)} \|f\|^{\frac{1}{m}}_{L_2(\mathbb{H}^d)}.$$

Thus,

$$c_\psi \|f\|^2_{W^{m,2}(\mathbb{H}^d)} \leq (4d)^m \sum_{n \in \mathbb{H}^d(\mathbb{Z})} \|f\psi_n\|^2_{W^{m,2}(\mathbb{H}^d)} + c^{(1)}_{m,\psi} c^2_m \|f\|^{2-\frac{2}{m}}_{W^{m,2}(\mathbb{H}^d)} \|f\|^{\frac{2}{m}}_{L_2(\mathbb{H}^d)}.$$

In other words,

$$(4d)^m \sum_{n \in \mathbb{H}^d(\mathbb{Z})} \|f\psi_n\|^2_{W^{m,2}(\mathbb{H}^d)} \geq c_\psi \|f\|^2_{W^{m,2}(\mathbb{H}^d)} - c^{(2)}_{m,\psi} \|f\|^{2-\frac{2}{m}}_{W^{m,2}(\mathbb{H}^d)} \|f\|^{\frac{2}{m}}_{L_2(\mathbb{H}^d)}.$$

On the other hand, we have

$$\sum_{n \in \mathbb{H}^d(\mathbb{Z})} \|f\psi_n\|^2_{W^{m,2}(\mathbb{H}^d)} \geq \sum_{n \in \mathbb{H}^d(\mathbb{Z})} \|f\psi_n\|^2_{L_2(\mathbb{H}^d)} \geq c_\psi \|f\|^2_{L_2(\mathbb{H}^d)}.$$

Choose $s > 0$ such that

$$c_\psi \lambda^2 - c^{(2)}_{m,\psi} \lambda^{2-\frac{2}{m}} \mu^{\frac{2}{m}} + sc_\psi \mu^2 \geq \frac{1}{2} c_\psi \lambda^2, \quad \lambda, \mu \geq 0.$$

We have

$$((4d)^m + s) \sum_{n \in \mathbb{H}^d(\mathbb{Z})} \|f\psi_n\|^2_{W^{m,2}(\mathbb{H}^d)}$$

$$\geq c_\psi \|f\|^2_{W^{m,2}(\mathbb{H}^d)} - c^{(2)}_{m,\psi} \|f\|^{2-\frac{2}{m}}_{W^{m,2}(\mathbb{H}^d)} \|f\|^{\frac{2}{m}}_{L_2(\mathbb{H}^d)} + sc_\psi \|f\|^2_{L_2(\mathbb{H}^d)}$$

$$\geq \frac{1}{2} c_\psi \|f\|^2_{W^{m,2}(\mathbb{H}^d)}.$$

Proof of Theorem 2.4.3 Let $\epsilon > 0$ be as in Lemma 2.4.4. Conjugating P with dilation, we may assume without loss of generality that each diam(supp(ψ)) $< \epsilon$.

Choose $l \in \mathbb{N}$ such that $\psi_{n_1} \psi_{n_2} = 0$ for $n_1, n_2 \in \delta_l(\mathbb{H}^d(\mathbb{Z}))$, $n_1 \neq n_2$. Let **n** be an element of the quotient space $\mathbb{H}^d(\mathbb{Z})/\delta_l(\mathbb{H}^d(\mathbb{Z}))$. We have $\psi_{n_1} \psi_{n_2} = 0$, $n_1, n_2 \in$ **n**, $n_1 \neq n_2$.

We have

$$\|Pf\|_{L_2(\mathbb{H}^d)} \geq \|M_{\sum_{n\in\mathbf{n}}\psi_n}Pf\|_{L_2(\mathbb{H}^d)}$$
$$\geq \|PM_{\sum_{n\in\mathbf{n}}\psi_n}f\|_{L_2(\mathbb{H}^d)} - \|[M_{\sum_{n\in\mathbf{n}}\psi_n},P]f\|_{L_2(\mathbb{H}^d)}.$$

Clearly,

$$PM_{\sum_{n\in\mathbf{n}}\psi_n}f = \sum_{n\in\mathbf{n}}P(f\psi_n),$$

where the summands on the right hand side are pairwise disjointly supported functions. Thus,

$$\|PM_{\sum_{n\in\mathbf{n}}\psi_n}f\|_{L_2(\mathbb{H}^d)} = \Big(\sum_{n\in\mathbf{n}}\|P(f\psi_n)\|^2_{L_2(\mathbb{H}^d)}\Big)^{\frac{1}{2}}$$

$$\stackrel{L.2.4.4}{\geq} \Big(\sum_{n\in\mathbf{n}}\big(c_P\|f\psi_n\|_{W^{m,2}(\mathbb{H}^d)} - \|f\psi_n\|_{L_2(\mathbb{H}^d)}\big)^2_+\Big)^{\frac{1}{2}}$$

$$\geq c_P\Big(\sum_{n\in\mathbf{n}}\|f\psi_n\|^2_{W^{m,2}(\mathbb{H}^d)}\Big)^{\frac{1}{2}} - \Big(\sum_{n\in\mathbf{n}}\|f\psi_n\|^2_{L_2(\mathbb{H}^d)}\Big)^{\frac{1}{2}},$$

$$\|[M_{\sum_{n\in\mathbf{n}}\psi_n},P]f\|_{L_2(\mathbb{H}^d)} \leq \|[M_{\sum_{n\in\mathbf{n}}\psi_n},P]\|_{W^{m-1,2}(\mathbb{H}^d)\to L_2(\mathbb{H}^d)}\|f\|_{W^{m-1,2}(\mathbb{H}^d)}.$$

Here, the right hand side is finite because, for the function $\theta = \sum_{n\in\mathbf{n}}\psi_n$, the function $X_w\theta$ is bounded for every word in the alphabet $\{1,\cdots,2d\}$ so that the differential operator $[M_\theta, P]$ of order $m-1$ has smooth coefficients constant outside some ball. Thus,

$$\|Pf\|_{L_2(\mathbb{H}^d)} \geq c_P\Big(\sum_{n\in\mathbf{n}}\|f\psi_n\|^2_{W^{m,2}(\mathbb{H}^d)}\Big)^{\frac{1}{2}} - c_P^{(1)}\|f\|_{W^{m-1,2}(\mathbb{H}^d)}.$$

Since the latter estimate holds for every $\mathbf{n}\in\mathbb{H}^d(\mathbb{Z})/\delta_l(\mathbb{H}^d(\mathbb{Z}))$, it follows that

$$\|Pf\|_{L_2(\mathbb{H}^d)} \geq \frac{c_P}{|\mathbb{H}^d(\mathbb{Z})/\delta_l(\mathbb{H}^d(\mathbb{Z}))|}\Big(\sum_{n\in\mathbb{H}^d(\mathbb{Z})}\|f\psi_n\|^2_{W^{m,2}(\mathbb{H}^d)}\Big)^{\frac{1}{2}} - c_P^{(1)}\|f\|_{W^{m-1,2}(\mathbb{H}^d)}.$$

It follows now from Lemma 2.4.6 that

$$\|Pf\|_{L_2(\mathbb{H}^d)} \geq \frac{c_P c_{m,\psi}}{|\mathbb{H}^d(\mathbb{Z})/\delta_l(\mathbb{H}^d(\mathbb{Z}))|^{\frac{1}{2}}}\|f\|_{W^{m,2}(\mathbb{H}^d)} - c_P^{(1)}\|f\|_{W^{m-1,2}(\mathbb{H}^d)}.$$

By complex interpolation, we have

$$\|f\|_{W^{m-1,2}(\mathbb{H}^d)} \le c_d \|f\|_{W^{m,2}(\mathbb{H}^d)}^{1-\frac{1}{m}} \|f\|_{L_2(\mathbb{H}^d)}^{\frac{1}{m}}.$$

Thus,

$$\|Pf\|_{L_2(\mathbb{H}^d)} \ge \frac{c_P c_{m,\psi}}{|\mathbb{H}^d(\mathbb{Z})/\delta_l(\mathbb{H}^d(\mathbb{Z}))|^{\frac{1}{2}}} \|f\|_{W^{m,2}(\mathbb{H}^d)} - c_P^{(1)} c_d \|f\|_{W^{m,2}(\mathbb{H}^d)}^{1-\frac{1}{m}} \|f\|_{L_2(\mathbb{H}^d)}^{\frac{1}{m}}.$$

Choose $s > 0$ such that

$$\frac{c_P c_{m,\psi}}{|\mathbb{H}^d(\mathbb{Z})/\delta_l(\mathbb{H}^d(\mathbb{Z}))|^{\frac{1}{2}}} \lambda - c_P^{(1)} c_d \lambda^{1-\frac{1}{m}} \mu^{\frac{1}{m}} + s\mu$$

$$\ge \frac{c_P c_{m,\psi}}{2|\mathbb{H}^d(\mathbb{Z})/\delta_l(\mathbb{H}^d(\mathbb{Z}))|^{\frac{1}{2}}} \lambda, \quad \lambda, \mu \ge 0.$$

We have

$$\|Pf\|_{L_2(\mathbb{H}^d)} + s\|f\|_{L_2(\mathbb{H}^d)}$$

$$\ge \frac{c_P c_{m,\psi}}{|\mathbb{H}^d(\mathbb{Z})/\delta_l(\mathbb{H}^d(\mathbb{Z}))|^{\frac{1}{2}}} \|f\|_{W^{m,2}(\mathbb{H}^d)}$$

$$- c_P^{(1)} c_d \|f\|_{W^{m,2}(\mathbb{H}^d)}^{\frac{1}{2}} \|f\|_{L_2(\mathbb{H}^d)}^{\frac{1}{2}} + s\|f\|_{L_2(\mathbb{H}^d)}$$

$$\ge \frac{c_P c_{m,\psi}}{2|\mathbb{H}^d(\mathbb{Z})/\delta_l(\mathbb{H}^d(\mathbb{Z}))|^{\frac{1}{2}}} \|f\|_{W^{m,2}(\mathbb{H}^d)}.$$

2.5 Self-Adjointness of Second Order Differential Operators

Definition 2.5.1 Let P be a differential operator of order 2 on $\mathbb{H}^d$ of the form:

$$P = -\sum_{k,l=1}^{2d} X_k M_{a_{kl}} X_l + \sum_{k=1}^{2d} M_{a_k} X_k + M_a,$$

where the coefficients are real-valued smooth functions on $\mathbb{H}^d$.

(i) We say that P is formally elliptic if there exists a constant $c > 0$ such that $(a_{kl})_{k,l=1}^{2d} \ge c$.

(ii) We say that P is formally self-adjoint if

$$\langle Pu, v\rangle_{L_2(\mathbb{H}^d)} = \langle u, Pv\rangle_{L_2(\mathbb{H}^d)}, \quad u, v \in W^{2,2}(\mathbb{H}^d).$$

(iii) We say that P is formally positive if

$$\langle Pu, u\rangle_{L_2(\mathbb{H}^d)} \geq 0, \quad u \in W^{2,2}(\mathbb{H}^d).$$

In Lemma 2.5.7, we will prove that formal ellipticity of a second order differential operator P yields uniform ellipticity of P^m for every $m \in \mathbb{N}$.

Theorem 2.5.2 *Let P be a differential operator of order 2 on $\mathbb{H}^d$. Suppose its coefficients are real-valued and constant outside some ball. If P is formally elliptic, formally self-adjoint and formally positive, then P is self-adjoint with domain $W^{2,2}(\mathbb{H}^d)$ and positive.*

The remaining part of this section is devoted to the proof of Theorem 2.5.2. So we assume that P is a formally elliptic differential operator of order 2 as in Theorem 2.5.2.

Lemma 2.5.3 *Let $c \in \mathbb{C}$. Every weak solution $u \in L_2(\mathbb{H}^d)$ of the equation $(P + c)u = 0$ is smooth on $\mathbb{H}^d$. The same assertion holds for the differential operator P on a connected open set $\Omega \subset \mathbb{H}^d$.*

Proof Let $A = (a_{kl})_{k,l=1}^{2d}$. By assumption, $A : \mathbb{H}^d \to \mathrm{GL}^+(2d, \mathbb{R})$ is smooth and constant outside some ball. So is $B = (b_{kl})_{k,l=1}^{2d} = A^{\frac{1}{2}}$. Set

$$Q_j = \sum_{k=1}^{2d} M_{b_{jk}} X_k, \quad 1 \leq j \leq 2d.$$

We, therefore, have

$$P = -\sum_{j=1}^{2d} Q_j^2 + Q_0,$$

where Q_0 is a differential operator of order 1 (in Heisenberg and, therefore, in Euclidean sense) with real-valued smooth coefficients.

Note that, for $1 \leq j_1, j_2 \leq 2d$,

$$[Q_{j_1}, Q_{j_2}] = M_{d_{j_1,j_2}} T + \sum_{k=1}^{2d} M_{b_{j_1,j_2,k}} X_k,$$

2.5 Self-Adjointness of Second Order Differential Operators

where

$$b_{j_1,j_2,k} = \sum_{l=1}^{2d} b_{j_1 l} \cdot X_l b_{j_2 k} - b_{j_2 l} \cdot X_l b_{j_1 k},$$

$$d_{j_1,j_2} = \sum_{k=1}^{d} b_{j_1,k} b_{j_2,k+d} - b_{j_1,k+d} b_{j_2,k}.$$

For every $p \in \mathbb{H}^d$, consider the operators

$$Q_{j,p} = \sum_{k=1}^{2d} b_{jk}(p) X_k, \quad 1 \le j \le 2d, \tag{2.5.1}$$

$$Q_{j_1,j_2,p} = d_{j_1,j_2}(p) T + \sum_{k=1}^{2d} b_{j_1,j_2,k}(p) X_k, \quad 1 \le j_1, j_2 \le 2d. \tag{2.5.2}$$

Since B is invertible, it follows from (2.5.1) that $X_k \in \text{span}(\{Q_{j,p}\}_{j=1}^{2d})$ for every $1 \le k \le 2d$ and

$$T = [X_1, X_{d+1}] \in \text{span}(\{[Q_{j_1,p}, Q_{j_2,p}]\}_{j_1,j_2=1}^{2d}).$$

By (2.5.2), we get

$$[Q_{j_1,p}, Q_{j_2,p}] = d_{j_1,j_2}(p) T \in Q_{j_1,j_2,p} + \text{span}(\{Q_{j,p}\}_{j=1}^{2d}),$$

and, therefore, $T \in \text{span}(\{Q_{j_1,j_2,p}\}_{j_1,j_2=1}^{2d}) + \text{span}(\{Q_{j,p}\}_{j=1}^{2d})$. It follows that, for every $p \in \mathbb{H}^d$, we have

$$\text{span}(\{Q_{j_1,j_2,p}\}_{j_1,j_2=1}^{2d}) + \text{span}(\{Q_{j,p}\}_{j=1}^{2d}) = \mathbb{H}^d.$$

This means that the operator $P + c$ satisfies the assumptions of Hörmander sum of squares theorem.

Note: in the paper [38] of Hörmander, the term of order 0 is allowed to be complex-valued, while terms of order 2 and 1 must be real-valued. This can be explicitly seen e.g. in the exposition of Hörmander theorem given in Section II.5 in [63]. Indeed, on p. 119 in [63] it is explicitly stated that c can be complex-valued.

The assertion follows by applying Hörmander sum of squares theorem to the operator $P + c$.

Recall the following result due to Williamson.

Theorem 2.5.4 *For every $A \in \text{GL}^+(2d, \mathbb{R})$, there exist $S \in \text{Sp}(2d, \mathbb{R})$ and diagonal $D \in \text{GL}^+(2d, \mathbb{R})$ with $d_{i,i} = d_{i+d,i+d}$ for $1 \le i \le d$ such that $A = S^* D S$.*

Here, $\{d_{i,i}\}_{i=1}^d$ are the positive eigenvalues of $\iota\Omega A$, while the negative eigenvalues of $\iota\Omega A$ are $\{-d_{i,i}\}_{i=1}^d$

Lemma 2.5.5 *Let P be a formally elliptic differential operator of order 2 with constant coefficients on $\mathbb{H}^d$. Every weak solution $u \in L_2(\mathbb{H}^d)$ of the equation $(P+\iota)u = f$ with $f \in L_2(\mathbb{H}^d)$ belongs to $W^{2,2}(\mathbb{H}^d)$.*

Proof We write $P = Q + R$, where

$$Q = -\sum_{k,l=1}^{2d} a_{kl} X_k X_l, \quad A = (a_{kl})_{k,l=1}^{2d} \in \mathrm{GL}^+(2d,\mathbb{R}),$$

and R is a differential operator of order 1 with constant coefficients.

By Theorem 2.5.4, $A = S^*DS$, where $S \in \mathrm{Sp}(2d,\mathbb{R})$ and D is diagonal such that $d_{i,i} = d_{i+d,i+d}$ for $1 \le i \le d$. Conjugating P with V_S and taking Lemma 2.1.7 into account, we may assume without loss of generality that $A = D$.

Using the Schrödinger representation (the analogue of Fourier transform for $\mathbb{H}^d$), we conclude that Q is self-adjoint (with domain $W^{2,2}(\mathbb{H}^d)$) and positive. It is of crucial importance that

$$W^{2m,2}(\mathbb{H}^d) = \{v \in \mathcal{S}'(\mathbb{H}^d) : (Q+1)^m v \in L_2(\mathbb{H}^d)\}.$$

Let $W^{-1,2}(\mathbb{H}^d)$ be the dual of $W^{1,2}(\mathbb{H}^d)$. It is easy to see that $W^{-1,2}(\mathbb{H}^d)$ consists of all distributions f on $\mathbb{H}^d$ such that $(Q+1)^{-\frac{1}{2}} f \in L_2(\mathbb{H}^d)$.

There exists a Hilbert inverse $S = (Q+\iota)^{-1}$. We clearly have $S \circ (Q+\iota) = \mathrm{id}$ on $\mathcal{S}'(\mathbb{H}^d)$. Since $R : L_2(\mathbb{H}^d) \to W^{-1,2}(\mathbb{H}^d)$, it follows that

$$(Q+\iota)u = f - Ru \in W^{-1,2}(\mathbb{H}^d).$$

Since $S : W^{-1,2}(\mathbb{H}^d) \to W^{1,2}(\mathbb{H}^d)$, it follows that

$$u = S((Q+\iota)u) = S(f - Ru) \in W^{1,2}(\mathbb{H}^d).$$

Since $R : W^{1,2}(\mathbb{H}^d) \to L_2(\mathbb{H}^d)$, it follows that

$$(Q+\iota)u = f - Ru \in L_2(\mathbb{H}^d).$$

Since $S : L_2(\mathbb{H}^d) \to W^{2,2}(\mathbb{H}^d)$, it follows that $u \in W^{2,2}(\mathbb{H}^d)$.

Lemma 2.5.6 *Let $A \in \mathrm{GL}^+(2d,\mathbb{R})$ and let D and S be as in Theorem 2.5.4. We have*

$$\|D\|_\infty \le \|A\|_\infty, \quad \|D^{-1}\|_\infty \le \|A^{-1}\|_\infty, \quad \|S\|_\infty, \|S^{-1}\|_\infty \le \|A^{-1}\|_\infty^{\frac{1}{2}} \|A\|_\infty^{\frac{1}{2}}.$$

2.5 Self-Adjointness of Second Order Differential Operators

Proof By Williamson theorem, $\{d_{k,k}\}_{k=1}^d$ are the positive eigenvalues of $\imath\Omega A$ (here Ω is defined by (1.3.5)). Hence,

$$\|D\|_\infty = \max_{1\le k\le d} d_{k,k} \le \|i\Omega A\|_\infty = \|A\|_\infty,$$

$$\|D^{-1}\|_\infty = \max_{1\le k\le d} d_{k,k}^{-1} \le \|(i\Omega A)^{-1}\|_\infty = \|A^{-1}\|_\infty.$$

Set $K = D^{\frac{1}{2}} S A^{-\frac{1}{2}}$. We have $S = D^{-\frac{1}{2}} K A^{\frac{1}{2}}$ and, therefore,

$$A = S^* D S = A^{\frac{1}{2}} K^* D^{-\frac{1}{2}} \cdot D \cdot D^{-\frac{1}{2}} K A^{\frac{1}{2}} = A^{\frac{1}{2}} \cdot K^* K \cdot A^{\frac{1}{2}}.$$

Since A is invertible, it follows that $K \in O(2d, \mathbb{R})$. Thus,

$$\|S\|_\infty \le \|A^{\frac{1}{2}}\|_\infty \|D^{-\frac{1}{2}}\|_\infty = \|A\|_\infty^{\frac{1}{2}} \|D^{-1}\|_\infty^{\frac{1}{2}} \le \|A^{-1}\|_\infty^{\frac{1}{2}} \|A\|_\infty^{\frac{1}{2}}.$$

Since $S \in \mathrm{Sp}(2d, \mathbb{R})$, it follows that $\|S^{-1}\|_\infty = \|S\|_\infty$. □

The following lemma shows that formal ellipticity of a second order differential operator P yields uniform ellipticity of P^m for every $m \in \mathbb{N}$. In this section, we need the assertion of the lemma only for $m = 1$. The full power of the lemma will be exploited in the next section.

Lemma 2.5.7 *Let* $A = (a_{k_1,k_2})_{k_1,k_2=1}^{2d} \in \mathrm{GL}^+(2d, \mathbb{R})$. *For every* $m \in \mathbb{N}$, *we have*

$$\left\|\left(\sum_{k_1,k_2=1}^{2d} a_{k_1,k_2} X_{k_1} X_{k_2}\right)^m \xi\right\|_{L_2(\mathbb{H}^d)} \ge c_{m,d} \|A^{-1}\|_\infty^{-2m} \|A\|_\infty^{-m} \|\Delta^m \xi\|_{L_2(\mathbb{H}^d)}$$

for every $\xi \in W^{2m,2}(\mathbb{H}^d)$.

Proof Let S and D be as in Theorem 2.5.4. Since $A = S^* D S$, it follows that $D = (S^{-1})^* A S^{-1}$. Set

$$B = \begin{pmatrix} S^{-1} & 0 \\ 0 & 1 \end{pmatrix} \in \mathrm{Aut}(\mathbb{H}^d).$$

Let $V_B : L_2(\mathbb{H}^d) \to L_2(\mathbb{H}^d)$ be the unitary operator defined by the formula $V_B \xi = \xi \circ B^*$.
Step 1: We claim that

$$\|V_B^{-1} \Delta^m V_B \eta\|_{L_2(\mathbb{H}^d)} \le c_{m,d} \|A^{-1}\|_\infty^m \|A\|_\infty^m \|\Delta^m \eta\|_{L_2(\mathbb{H}^d)}, \quad \eta \in W^{2m,2}(\mathbb{H}^d).$$

By Lemma 2.1.7, we have

$$V_B^{-1} \Delta V_B = \sum_{k_1,k_2=1}^{2d} (B^* B)_{k_1,k_2} X_{k_1} X_{k_2}.$$

Thus,
$$V_B^{-1}\Delta^m V_B = \sum_{1\leq k_1,\cdots,k_{2m}\leq 2d}\prod_{l=1}^{m}(B^*B)_{k_{2l-1},k_{2l}}\cdot\prod_{l=1}^{2m}X_{k_l}.$$

By triangle inequality,
$$\|V_B^{-1}\Delta^m V_B\eta\|_{L_2(\mathbb{H}^d)} \leq \sum_{1\leq k_1,\cdots,k_{2m}\leq 2d}\prod_{l=1}^{m}|(B^*B)_{k_{2l-1},k_{2l}}|\cdot\left\|\left(\prod_{l=1}^{2m}X_{k_l}\right)\eta\right\|_{L_2(\mathbb{H}^d)}$$

$$\leq \|B\|_\infty^{2m}\cdot\sum_{1\leq k_1,\cdots,k_{2m}\leq 2d}\left\|\left(\prod_{l=1}^{2m}X_{k_l}\right)\eta\right\|_{L_2(\mathbb{H}^d)}$$

$$\stackrel{P.2.1.9}{\leq} c_{m,d}\|B\|_\infty^{2m}\|\Delta^m\eta\|_{L_2(\mathbb{H}^d)}.$$

The claim follows now from Lemma 2.5.6.
Step 2: We claim that

$$\|V_B^{-1}(\sum_{k_1,k_2=1}^{2d}a_{k_1,k_2}X_{k_1}X_{k_2})^m V_B\eta\|_{L_2(\mathbb{H}^d)}$$

$$\geq \|A^{-1}\|_\infty^{-m}\|\Delta^m\eta\|_{L_2(\mathbb{H}^d)},\quad \eta\in W^{2m,2}(\mathbb{H}^d).$$

By Lemma 2.1.7, we have

$$V_B^{-1}\left(\sum_{k_1,k_2=1}^{2d}a_{k_1,k_2}X_{k_1}X_{k_2}\right)V_B = \sum_{k_1,k_2=1}^{2d}a_{k_1,k_2}V_B^{-1}X_{k_1}V_B\cdot V_B^{-1}X_{k_2}V_B$$

$$= \sum_{k_1,k_2=1}^{2d}a_{k_1,k_2}\sum_{j_1,j_2=1}^{2d}b_{k_1,j_1}b_{k_2,j_2}X_{j_1}X_{j_2}$$

$$= \sum_{j_1,j_2=1}^{2d}\left(\sum_{k_1,k_2=1}^{2d}b_{k_1,j_1}a_{k_1,k_2}b_{k_2,j_2}\right)X_{j_1}X_{j_2}$$

$$= \sum_{k_1,k_2=1}^{2d}((S^{-1})^*AS^{-1})_{k_1,k_2}X_{k_1}X_{k_2}$$

$$= \sum_{k=1}^{d}\lambda_k(X_k^2+X_{k+d}^2).$$

2.5 Self-Adjointness of Second Order Differential Operators

It is immediate that

$$\|V_B^{-1}(\sum_{k_1,k_2=1}^{2d} a_{k_1,k_2} X_{k_1} X_{k_2})^m V_B \eta\|_{L_2(\mathbb{H}^d)} = \|(\sum_{k=1}^d \lambda_k (X_k^2 + X_{k+d}^2))^m \eta\|_{L_2(\mathbb{H}^d)}$$

$$\geq (\min_{1\leq k\leq d} \lambda_k)^m \|\Delta^m \eta\|_{L_2(\mathbb{H}^d)}.$$

The claim follows now from Lemma 2.5.6.

Step 3: Now we are ready to complete the proof of the lemma.

It follows from Lemma 2.1.7 that the unitary operators $V_B, V_B^{-1} : L_2(\mathbb{H}^d) \to L_2(\mathbb{H}^d)$ map $W^{2m,2}(\mathbb{H}^d)$ into itself.

Fix $\xi \in W^{2m,2}(\mathbb{H}^d)$ and set $\eta = V_B^{-1} \xi \in W^{2m,2}(\mathbb{H}^d)$. By Step 1, we have

$$\|\Delta^m \xi\|_{L_2(\mathbb{H}^d)} = \|V_B^{-1} \Delta^m \xi\|_{L_2(\mathbb{H}^d)}$$
$$= \|V_B^{-1} \Delta^m V_B \eta\|_{L_2(\mathbb{H}^d)}$$
$$\leq c_{m,d} \|A^{-1}\|_\infty^m \|A\|_\infty^m \|\Delta^m \eta\|_{L_2(\mathbb{H}^d)}.$$

By Step 2, we have

$$\|(\sum_{k_1,k_2=1}^{2d} a_{k_1,k_2} X_{k_1} X_{k_2})^m \xi\|_{L_2(\mathbb{H}^d)} = \|V_B^{-1}(\sum_{k_1,k_2=1}^{2d} a_{k_1,k_2} X_{k_1} X_{k_2})^m V_B \eta\|_{L_2(\mathbb{H}^d)}$$

$$\geq \|A^{-1}\|_\infty^{-m} \|\Delta^m \eta\|_{L_2(\mathbb{H}^d)}.$$

The claim follows by combining these inequalities.

We observe the following easy fact.

Fact 2.5.8 *Let self-adjoint positive operators B_1 and B_2 have the same domain. Suppose $\ker(B_2) = 0$ and $\overline{\mathrm{im}(B_2)} = H$. If $\|B_1 \xi\| \leq \|B_2 \xi\|$ for every $\xi \in \mathrm{dom}(B_1)$, then $B_1 B_2^{-1}$ extends to a contraction on H.*

Lemma 2.5.9 *For every $A = (a_{k_1,k_2})_{k_1,k_2=1}^{2d} \in \mathrm{GL}^+(2d, \mathbb{R})$, the operator*

$$\Delta^m (\sum_{k_1,k_2=1}^{2d} a_{k_1,k_2} X_{k_1} X_{k_2})^{-m}$$

extends to a bounded operator in $L_2(\mathbb{H}^d)$ with the following norm estimate

$$\left\|\Delta^m (\sum_{k_1,k_2=1}^{2d} a_{k_1,k_2} X_{k_1} X_{k_2})^{-m}\right\|_\infty \leq c_{m,d} \|A^{-1}\|_\infty^{2m} \|A\|_\infty^m. \qquad (2.5.3)$$

Proof We apply Fact 2.5.8 to the operators

$$B_1 = \Delta^m, \quad B_2 = \left(-\sum_{k_1,k_2=1}^{2d} a_{k_1,k_2} X_{k_1} X_{k_2}\right)^m$$

on $\mathrm{dom}((\sum_{k_1,k_2=1}^{2d} a_{k_1,k_2} X_{k_1} X_{k_2})^m) = \mathrm{dom}(\Delta^m) = W^{2m,2}(\mathbb{H}^d)$. It is clear that B_1 and B_2 are formally self-adjoint, formally positive, have the same domain. By Proposition 2.4.3, they are closed. Since B_1 and B_2 have constant coefficients (cf. Lemma 2.5.5), they are self-adjoint and positive. By Lemma 2.5.7, the operator $B_2 : W^{2m,2}(\mathbb{H}^d) \to L_2(\mathbb{H}^d)$ is injective with dense image. So the desired assertion follows immediately from Fact 2.5.8.

Proof of Theorem 2.5.2 Applying Lemma 2.5.7 with $m = 1$, we obtain that P is uniformly elliptic. It follows from Theorem 2.4.3 that $P : W^{2,2}(\mathbb{H}^d) \to L_2(\mathbb{H}^d)$ is closed.

To show self-adjointness, it suffices (see Theorem VIII.3 in [57]) to establish that every $u \in \mathrm{dom}(P^*)$ with $(P^* + \iota)u = 0$ (or with $(P^* - \iota)u = 0$) is trivial.

Denote by P' the same differential operator P acting on distributions. We have $P^* = P'|_{\mathrm{dom}(P^*)}$. If $u \in \mathrm{dom}(P^*)$ with $(P^* + \iota)u = 0$, then $u \in L_2(\mathbb{H}^d)$ and $(P' + \iota)u = 0$. In other words, u is a weak solution of the equation $(P + \iota)u = 0$. By Lemma 2.5.3, u is smooth.

Recall that coefficients of P are constant outside some ball. Let P_0 be a differential operator of order 2 with these constant coefficients. We have that $P_0 - P$ is a differential operator of order 2 whose coefficients are compactly supported. Since u is smooth, it follows that $(P_0 - P)u \in C_c^\infty(\mathbb{H}^d)$. Hence,

$$(P_0 + \iota)u = (P + \iota)u + (P_0 - P)u = (P_0 - P)u \in C_c^\infty(\mathbb{H}^d) \subset L_2(\mathbb{H}^d).$$

It follows from Lemma 2.5.5 that $u \in W^{2,2}(\mathbb{H}^d)$.

Since P is formally self-adjoint, it follows that

$$\langle Pu, u \rangle_{L_2(\mathbb{H}^d)} = \langle u, Pu \rangle_{L_2(\mathbb{H}^d)}.$$

Noting that $Pu = -\iota u$, we obtain

$$-\iota \|u\|^2_{L_2(\mathbb{H}^d)} = \langle -\iota u, u \rangle_{L_2(\mathbb{H}^d)} = \langle u, -\iota u \rangle_{L_2(\mathbb{H}^d)} = \iota \|u\|^2_{L_2(\mathbb{H}^d)}.$$

Thus, $\|u\|_{L_2(\mathbb{H}^d)} = 0$ and $u = 0$.

2.6 Local Connes Trace Formula

This section is devoted to the Connes trace formula for an elliptic, self-adjoint and positive differential operator of order 2 on the Heisenberg group.

Theorem 2.6.1 *Let P be an elliptic, self-adjoint and positive differential operator of order 2 on $\mathbb{H}^d$. Suppose the coefficients of P are smooth and constant outside some ball. For every compactly supported $S \in \Pi$, we have*

$$S(1+P)^{-d-1} \in \mathcal{L}_{1,\infty},$$

$$\varphi(S(1+P)^{-d-1}) = c_d \Big(\int_{\mathbb{H}^d} \otimes \tau \Big) \Big(\mathrm{sym}(S) \cdot e^{\sum_{k_1,k_2=1}^{2d} a_{k_1 k_2} \otimes X_{k_1} X_{k_2}} \Big).$$

Here, the principal part of P is $-\sum_{k_1,k_2=1}^{2d} X_{k_1} M_{a_{k_1 k_2}} X_{k_2}$.

The rest of this section is devoted to the proof of Theorem 2.6.1. We will start with a technical lemma.

Lemma 2.6.2 *Suppose that $x \in B(L_2(\mathbb{H}^d))$ is a boundedly invertible operator of the form $x = y + r$, where:*

(i) $y \in \Pi$ is such that $\mathrm{sym}(y)$ is boundedly invertible;
(ii) $r \in B(L_2(\mathbb{H}^d))$ is such that $M_\phi r$ and $r M_\phi$ are compact for some $\phi \in C_c^\infty(\mathbb{H}^d)$.

Under those assumptions, we have $M_\phi x^{-1} \in \Pi$ and

$$\mathrm{sym}(M_\phi x^{-1}) = (\phi \otimes 1) \cdot \mathrm{sym}(y)^{-1}.$$

Proof
Step 1: Let P be a polynomial in 2 non-commuting arguments. We claim that $M_\phi P(x, x^*) \in \Pi$.

We prove the assertion by induction on $\deg(P)$. The base of induction (i.e., the assertion for constant P) is obvious. Let us establish the step of induction (i.e., let us assume the assertion for $\deg(P) = m$ and let us prove it for $\deg(P) = m+1$).

Without loss of generality, P is monomial. We either have $P(u,v) = uQ(u,v)$ or $P(u,v) = vQ(u,v)$, where $\deg(Q) = m$. In the first case, we write

$$M_\phi P(x, x^*) = y \cdot M_\phi Q(x, x^*) + [M_\phi, y] \cdot Q(x, x^*) + M_\phi r \cdot Q(x, x^*).$$

In the second case, we write

$$M_\phi P(x, x^*) = y^* \cdot M_\phi Q(x, x^*) + [M_\phi, y^*] \cdot Q(x, x^*) + M_\phi r^* \cdot Q(x, x^*).$$

Since $M_\phi r$ and $r M_\phi$ are compact, it follows that the third summand is compact. Since $y \in \Pi$, it follows from Lemma 2.2.1 that the second summand is compact. By Theorem 1.3.2, second and third summands are in Π and their symbols are zeroes.

By inductive assumption, we have $M_\phi Q(x, x^*) \in \Pi$. Since $y \in \Pi$, it follows that $M_\phi P(x, x^*) \in \Pi$. This establishes the step of induction and, hence, yields the claim.

Step 2: We now claim that $M_\phi x^{-1} \in \Pi$.

Since x is boundedly invertible, it follows that $x^{-1} \in C^*(x)$. Hence, we can choose a sequence $\{P_n\}_{n\geq 0}$ of polynomials in 2 non-commuting arguments such that

$$P_n(x, x^*) \to x^{-1}$$

in the uniform norm. By Step 1, we have $M_\phi P_n(x, x^*) \in \Pi$ for every $n \geq 0$. Since Π is uniform norm-closed, it follows that $M_\phi x^{-1} \in \Pi$.

Step 3: We now claim that

$$\mathrm{sym}(M_\phi x^{-1}) = (\phi \otimes 1) \cdot \mathrm{sym}(y)^{-1}.$$

Note that

$$[M_\phi, x^{-1}] = x^{-1} \cdot [x, M_\phi] \cdot x^{-1} = x^{-1} \cdot [y, M_\phi] \cdot x^{-1} + x^{-1} \cdot [r, M_\phi] \cdot x^{-1}.$$

Since $y \in \Pi$, it follows from Lemma 2.2.1 that the first summand on the right hand side is compact. The operators $r M_\phi$ and $M_\phi r$ are compact and, hence, so is the second summand on the right hand side. Thus, $[M_\phi, x^{-1}]$ is compact.

Now,

$$M_\phi x^{-1} \cdot y - M_\phi = [M_\phi, x^{-1}] \cdot y - x^{-1} \cdot M_\phi r + x^{-1} \cdot [M_\phi, x]$$

is compact. By definition, this means

$$\mathrm{sym}(M_\phi x^{-1}) \cdot \mathrm{sym}(y) = \mathrm{sym}(M_\phi x^{-1} \cdot y) = \mathrm{sym}(M_\phi).$$

Since $\mathrm{sym}(y)$ is boundedly invertible, the assertion follows. □

Lemma 2.6.3 *For every $m \in \mathbb{N}$ and for every $\{k_l\}_{l=1}^{2m}$, we have*

$$\Delta^{-m} \prod_{l=1}^{2m} X_{k_l} \in C^*(\{R_k\}_{k=1}^{2d}).$$

Proof Let us prove the assertion by induction on m.

The key fact we are using is: for every $x \in \mathcal{K}(L_2(\mathbb{R}^d))$ and for every $\xi \in \mathbb{C}^2$, the operator $\pi_{\mathrm{red}}(x \otimes \xi) \in C^*(\{R_k\}_{k=1}^{2d})$. This fact follows from Lemma 5.2 in [29].

2.6 Local Connes Trace Formula

Firstly, let us prove the base of induction (i.e., $m = 1$). Assume for simplicity of notations that $1 \leq k_1, k_2 \leq d$. We have

$$\Delta^{-1} X_{k_1} X_{k_2} = \pi_{\text{red}}\Big(H_d^{-1} p_{k_1} p_{k_2} \otimes (1, 1)\Big).$$

It is immediate that (e.g. by considering the operators in Hermite basis)

$$H_d^{-1} p_{k_1} p_{k_2} - H_d^{-\frac{1}{2}} p_{k_1} p_{k_2} H_d^{-\frac{1}{2}}$$

is compact. Therefore, we have

$$\pi_{\text{red}}\Big(\big(H_d^{-1} p_{k_1} p_{k_2} - H_d^{-\frac{1}{2}} p_{k_1} p_{k_2} H_d^{-\frac{1}{2}}\big) \otimes (1, 1)\Big) \in C^*(\{R_k\}_{k=1}^{2d}).$$

On the other hand, we have

$$\pi_{\text{red}}\Big(\big(H_d^{-\frac{1}{2}} p_{k_1} p_{k_2} H_d^{-\frac{1}{2}}\big) \otimes (1, 1)\Big) = R_{k_1}^* R_{k_2} \in C^*(\{R_k\}_{k=1}^{2d}).$$

Combining these inclusions, we establish the base of induction.

Now, let us prove the step of induction. Suppose the assertion holds for m and let us prove it for $m + 1$. Assume for simplicity of notations that $1 \leq k_1, k_2, \cdots, k_{2m+2} \leq d$. We have

$$\Delta^{-m-1} \prod_{l=1}^{2m+2} X_{k_l} = \pi_{\text{red}}\Big(H_d^{-m-1} \prod_{l=1}^{2m+2} p_{k_l} \otimes (1, 1)\Big).$$

It is immediate that

$$H_d^{-m-1} \prod_{l=1}^{2m+2} p_{k_l} - H_d^{-m}\Big(\prod_{l=1}^{2m+2} p_{k_l}\Big) H_d^{-1}$$

is compact. Therefore, we have

$$\pi_{\text{red}}\Big(\big(H_d^{-m-1} \prod_{l=1}^{2m+2} p_{k_l} - H_d^{-m}\big(\prod_{l=1}^{2m+2} p_{k_l}\big) H_d^{-1}\big) \otimes (1, 1)\Big) \in C^*(\{R_k\}_{k=1}^{2d}).$$

On the other hand, we have

$$\pi_{\text{red}}\Big(\big(H_d^{-m}\big(\prod_{l=1}^{2m+2} p_{k_l}\big) H_d^{-1}\big) \otimes (1, 1)\Big) = \Delta^{-m} \prod_{l=1}^{2m} X_{k_l} \cdot \Big(-\Delta^{-1} X_{k_2} X_{k_1}\Big)^*.$$

The first factor on the right hand side belongs to $C^*(\{R_k\}_{k=1}^{2d})$ by inductive assumption. The second factor on the right hand side belongs to $C^*(\{R_k\}_{k=1}^{2d})$ by the base of induction. Thus,

$$\pi_{\text{red}}\left(\left(H_d^{-m}\left(\prod_{l=1}^{2m+2} p_{k_l}\right)H_d^{-1}\right) \otimes (1,1)\right) \in C^*(\{R_k\}_{k=1}^{2d}).$$

Combining these inclusions, we establish the step of induction.

Let $A = (a_{k_1 k_2})_{k_1,k_2=1}^{2d} \in C^\infty(\mathbb{H}^d, \text{GL}^+(2d, \mathbb{R}))$. Suppose A is constant outside some ball. By Lemma 2.6.3, the formula

$$S = (1 \otimes \Delta)^{-d-1}\left(-\sum_{k_1,k_2=1}^{2d} a_{k_1 k_2} \otimes X_{k_1} X_{k_2}\right)^{d+1}$$

defines an element of $L_\infty(\mathbb{H}^d) \bar{\otimes} \text{VN}_{\text{hom}}(\mathbb{H}^d)$.

Lemma 2.6.4 *The element S is invertible in $L_\infty(\mathbb{H}^d) \bar{\otimes} \text{VN}_{\text{hom}}(\mathbb{H}^d)$.*

Proof By Lemma 2.5.9, for any $p \in \mathbb{H}^d$, the operator

$$T(p) = \Delta^{d+1}\left(-\sum_{k_1,k_2=1}^{2d} a_{k_1,k_2}(p) X_{k_1} X_{k_2}\right)^{-d-1}$$

is a bounded operator in $L_2(\mathbb{H}^d)$, and

$$\sup_{p \in \mathbb{H}^d} \|T(p)\|_\infty \leq c_{d+1,d} \sup_{p \in \mathbb{H}^d} \|A(p)^{-1}\|_\infty^{2d+2} \|A(p)\|_\infty^{d+1} < \infty.$$

Therefore, $T = (1 \otimes \Delta)^{d+1}\left(-\sum_{k_1,k_2=1}^{2d} a_{k_1 k_2} \otimes X_{k_1} X_{k_2}\right)^{d+1}$ is a well defined element of $L_\infty(\mathbb{H}^d) \bar{\otimes} \text{VN}_{\text{hom}}(\mathbb{H}^d)$ which is the inverse of S.

Lemma 2.6.5 *Let P be a formally elliptic, self-adjoint and positive differential operator of order 2. Suppose the coefficients of P are smooth and constant outside some ball. We have*

$$M_\phi (1+P)^{-d-1}(1+\Delta)^{d+1} \in \Pi, \quad \phi \in C_c^\infty(\mathbb{H}^d).$$

If principal part of P is $-\sum_{k_1,k_2=1}^{2d} X_{k_1} M_{a_{k_1 k_2}} X_{k_2}$, then

$$\text{sym}\left(M_\phi (1+P)^{-d-1}(1+\Delta)^{d+1}\right) = (\phi \otimes 1) \cdot S_P^{-1},$$

2.6 Local Connes Trace Formula

where

$$S_P = (1 \otimes \Delta)^{-d-1}\Big(-\sum_{k_1,k_2=1}^{2d} a_{k_1 k_2} \otimes X_{k_1} X_{k_2}\Big)^{d+1}.$$

Proof Denote for brevity

$$Q = (-1)^{d+1} \sum_{1 \leq k_1,\cdots,k_{2d+2} \leq 2d} \prod_{l=1}^{2d+2} X_{k_l} \cdot \prod_{l=1}^{d+1} M_{a_{k_{2l-1} k_{2l}}},$$

$$R = (1+P)^{d+1} - Q,$$

and note that R is a differential operator of order $2d + 1$.

Now we want to apply Lemma 2.6.2 to the operators

$$x = (1+\Delta)^{-d-1}(1+P)^{d+1}, \quad y = \Delta^{-d-1} Q, \quad r = x - y.$$

Let us check the conditions of this lemma.

Since $(1+P)^{d+1}$ is a differential operator of order $2d + 2$, it follows from Lemma 2.6.3 that $x \in B(L_2(\mathbb{H}^d))$. By Lemma 2.6.3, we have $y \in \Pi$. Thus, $r = x - y \in B(L_2(\mathbb{H}^d))$.

It follows from Lemma 2.5.9 that $(1+P)^{d+1}$ is uniformly elliptic. Applying Theorem 2.4.3 to the operator $(1+P)^{d+1}$, we conclude that

$$(1+P)^{-d-1} : L_2(\mathbb{H}^d) \to W^{2d+2,2}(\mathbb{H}^d)$$

is bounded. It follows that

$$(x^{-1})^* = (1+\Delta)^{d+1}(1+P)^{-d-1}$$

is bounded on $L_2(\mathbb{H}^d)$. Hence, x is boundedly invertible.

It is immediate that

$$\text{sym}(y) = (-1)^{d+1} \sum_{1 \leq k_1,\cdots,k_{2d+2} \leq 2d} \Big(\prod_{l=1}^{d+1} a_{k_{2l-1} k_{2l}}\Big) \otimes \Big(\Delta^{-d-1} \prod_{l=1}^{2d+2} X_{k_l}\Big) = S_P.$$

By Lemma 2.6.4, sym(y) is boundedly invertible.

It remains to show that $M_\phi r, r M_\phi \in \mathcal{K}(L_2(\mathbb{H}^d))$. By definition, $r = r_1 + r_2$, where

$$r_1 = (1+\Delta)^{-d-1} R, \quad r_2 = \Big((1+\Delta)^{-d-1} - \Delta^{-d-1}\Big) Q.$$

Since R is a differential operator of order $2d+1$, it follows that the operator $(1+\Delta)^{-d-\frac{1}{2}}R$ is bounded on $L_2(\mathbb{H}^d)$. Thus,

$$M_\phi r_1 = M_\phi \Delta^{-\frac{1}{2}} \cdot \Delta^{-d-\frac{1}{2}}R$$

is compact. Choose $\psi \in C_c^\infty(\mathbb{H}^d)$ such that $\phi\psi = \phi$. Noting that R is a differential operator, we obtain $RM_\phi = M_\psi RM_\phi$. Thus,

$$r_1 M_\phi = (1+\Delta)^{-d-1} M_\psi R M_\phi = (1+\Delta)^{-d-1} M_\psi (1+\Delta)^{d+\frac{1}{2}} \cdot (1+\Delta)^{-d-\frac{1}{2}} R M_\phi$$

is compact.

Clearly,

$$M_\phi r_2 = M_\phi \left(\left(\frac{\Delta}{1+\Delta} \right)^{d+1} - 1 \right) \cdot z$$

is compact. Since Q is a differential operator, it follows that $QM_\phi = M_\psi Q M_\phi$. Since Q is a differential operator of order $2d+2$, it follows that $(1+\Delta)^{-d-1}Q$ is bounded on $L_2(\mathbb{H}^d)$. Thus,

$$r_2 M_\phi = \left((1+\Delta)^{-d-1} - \Delta^{-d-1} \right) M_\psi Q M_\phi$$
$$= \left((1+\Delta)^{-d-1} - \Delta^{-d-1} \right) M_\psi (1+\Delta)^{d+1} \cdot (1+\Delta)^{-d-1} Q M_\phi$$

is compact.

We established that the operators $M_\phi r_1$, $M_\phi r_2$, $r_1 M_\phi$ and $r_2 M_\phi$ are compact. Since $r = r_1 + r_2$, the operators rM_ϕ and $M_\phi r$ are compact.

Thus, the conditions of Lemma 2.6.2 hold true. The assertion of the lemma follows immediately from Lemma 2.6.2.

Recall that the group von Neumann algebra $\mathrm{VN}(\mathbb{H}^d)$ is $*$-isomorphic (the $*$-isomorphism is denoted by π) to $B(L_2(\mathbb{R}^d))\bar{\otimes}L_\infty(\mathbb{R},|s|^d ds)$ (see Proposition 2.1.4). By Proposition 2.1.4, the semifinite trace τ on $\mathrm{VN}(\mathbb{H}^d)$ defined in Definition 2.1.2 can be written as

$$\tau(\pi(x \otimes f)) = \mathrm{Tr}(x) \cdot \int_\mathbb{R} f(s)|s|^d ds. \tag{2.6.1}$$

The group homogeneous von Neumann algebra $\mathrm{VN}_{\mathrm{hom}}(\mathbb{H}^d)$ is $*$-isomorphic (the $*$-isomorphism is denoted by π_{red}) to $B(L_2(\mathbb{R}^d)) \otimes \mathbb{C}^2$ (see Sect. 2.1.5). There is a natural semifinite trace τ_{hom} on this von Neumann algebra given by (2.1.6).

2.6 Local Connes Trace Formula

Lemma 2.6.6 *Let P be an elliptic, self-adjoint and positive differential operator of order 2. Suppose the coefficients of P are smooth and constant outside some ball. For every continuous normalised trace φ on $\mathcal{L}_{1,\infty}$ and for every compactly supported $S \in \Pi$, we have*

$$S(1+P)^{-d-1} \in \mathcal{L}_{1,\infty},$$

$$\varphi(S(1+P)^{-d-1}) = c_d \Big(\int_{\mathbb{H}^d} \otimes \tau_{\text{hom}} \Big) \Big(\text{sym}(S) \cdot \Big(\frac{1 \otimes |T|}{-\sum_{k_1,k_2=1}^{2d} a_{k_1 k_2} \otimes X_{k_1} X_{k_2}} \Big)^{d+1} \Big).$$

Here, the principal part of P is $-\sum_{k_1,k_2=1}^{2d} X_{k_1} M_{a_{k_1,k_2}} X_{k_2}$.

Proof Fix $\phi \in C_c^\infty(\mathbb{H}^d)$ such that $S = M_\phi S = S M_\phi$. By the tracial property, we have

$$\varphi(S(1+P)^{-d-1}) = \varphi(M_\phi S M_\phi (1+P)^{-d-1}) = \varphi(S M_\phi (1+P)^{-d-1} M_\phi).$$

Clearly,

$$S M_\phi (1+P)^{-d-1} M_\phi = S S' \cdot (1+\Delta)^{-d-1} M_\phi,$$

where

$$S' = M_\phi (1+P)^{-d-1} (1+\Delta)^{d+1}.$$

By Lemma 2.6.5, we have $S' \in \Pi$. We write

$$S S' \cdot (1+\Delta)^{-d-1} M_\phi = S S' M_\phi \cdot (1+\Delta)^{-d-1} + S S' \cdot [(1+\Delta)^{-d-1}, M_\phi].$$

The commutator on the right hand side belongs to $\mathcal{L}_{\frac{2d+2}{2d+3},\infty}$. Hence,

$$\varphi(S(1+P)^{-d-1}) = \varphi(S S' \cdot (1+\Delta)^{-d-1} M_\phi) = \varphi(S S' M_\phi \cdot (1+\Delta)^{-d-1}).$$

Since $S' \in \Pi$, it follows that $S S' M_\phi \in \Pi$ is compactly supported. By Lemma 2.3.5, we have

$$\varphi(S S' M_\phi \cdot (1+\Delta)^{-d-1}) = c_d \Big(\int_{\mathbb{H}^d} \otimes \tau_{\text{hom}} \Big) \Big(\text{sym}(S S' M_\phi) \cdot \Big(1 \otimes \frac{|T|}{\Delta}\Big)^{d+1} \Big).$$

By Lemma 2.6.5, we have

$$\text{sym}(S S' M_\phi) = \text{sym}(S) \cdot \text{sym}(S') \cdot (\phi \otimes 1) = \text{sym}(S) \cdot S_P^{-1} \cdot (\phi^2 \otimes 1) = \text{sym}(S) \cdot S_P^{-1}.$$

Note that

$$S_P^{-1}\Big(1 \otimes \frac{|T|}{\Delta}\Big)^{d+1} = (-1)^{d+1}\Big(\sum_{k_1,k_2=1}^{2d} a_{k_1 k_2} \otimes X_{k_1} X_{k_2}\Big)^{-d-1} \cdot (1 \otimes |T|^{d+1})$$

$$= \Big(\frac{1 \otimes |T|}{-\sum_{k_1,k_2=1}^{2d} a_{k_1 k_2} \otimes X_{k_1} X_{k_2}}\Big)^{d+1}.$$

This completes the proof.

The remaining argument in Theorem 2.6.1 is similar to the one in Theorem 1.4.1.

Proof of Theorem 2.6.1 We have

$$\Big(\int_{\mathbb{H}^d} \otimes \tau_{\text{hom}}\Big)\Big(\text{sym}(S) \cdot \Big(\frac{1 \otimes |T|}{-\sum_{k_1,k_2=1}^{2d} a_{k_1 k_2} \otimes X_{k_1} X_{k_2}}\Big)^{d+1}\Big)$$

$$= \int_{\mathbb{H}^d} \tau_{\text{hom}}\Big(\text{sym}(S)(p) \cdot \Big(\frac{1 \otimes |T|}{-\sum_{k_1,k_2=1}^{2d} a_{k_1 k_2}(p) X_{k_1} X_{k_2}}\Big)^{d+1}\Big) dp.$$

Applying Lemma 2.3.6 with

$$x = \text{sym}(S)(p), \quad y = -\frac{\sum_{k_1,k_2=1}^{2d} a_{k_1 k_2}(p) X_{k_1} X_{k_2}}{|T|},$$

we obtain

$$\tau_{\text{hom}}\Big(\text{sym}(S)(p) \cdot \Big(\frac{1 \otimes |T|}{-\sum_{k_1,k_2=1}^{2d} a_{k_1 k_2}(p) X_{k_1} X_{k_2}}\Big)^{d+1}\Big)$$

$$= c_d \tau\Big(\text{sym}(S)(p) \cdot e^{\sum_{k_1,k_2=1}^{2d} a_{k_1 k_2}(p) X_{k_1} X_{k_2}}\Big), \quad p \in \mathbb{H}^d.$$

Thus,

$$\Big(\int_{\mathbb{H}^d} \otimes \tau_{\text{hom}}\Big)\Big(\text{sym}(S) \cdot \Big(\frac{1 \otimes |T|}{-\sum_{k_1,k_2=1}^{2d} a_{k_1 k_2} \otimes X_{k_1} X_{k_2}}\Big)^{d+1}\Big)$$

$$= c_d \int_{\mathbb{H}^d} \tau\Big(\text{sym}(S)(p) \cdot e^{\sum_{k_1,k_2=1}^{2d} a_{k_1 k_2}(p) X_{k_1} X_{k_2}}\Big) dp$$

$$= c_d \Big(\int_{\mathbb{H}^d} \otimes \tau\Big)\Big(\text{sym}(S) \cdot e^{\sum_{k_1,k_2=1}^{2d} a_{k_1 k_2} \otimes X_{k_1} X_{k_2}}\Big).$$

The assertion follows now from Lemma 2.6.6.

Chapter 3
Equivariance of the Principal Symbol Under Heisenberg Diffeomorphisms

In this chapter, we give a positive answer to a Heisenberg analogue of Question 1.1.2 on the properties of the C^*-algebra Π, that is, we show that the C^*-algebra Π is invariant under the natural unitary action of the group of Heisenberg diffeomorphisms on $B(L_2(\mathbb{R}^d))$ satisfying some additional assumptions and the $*$-homomorphism sym behaves equivariantly under this action.

3.1 The Action on the Codomain of the Principal Symbol Map

In this section, we define a relevant action of Heisenberg diffeomorphisms on the algebra $C(\Omega) \otimes_{\min} C^*(\{R_k\}_{k=1}^{2d})$, the codomain (and, due to surjectivity, the image) of the principal symbol map.

First, we prove that the action of $\mathrm{Aut}(\mathbb{H}^d)$ on $C^*(\{R_k\}_{k=1}^{2d})$ is well defined.

Theorem 3.1.1 *Let* $A \in \mathrm{Aut}(\mathbb{H}^d)$. *The $*$-automorphism π_A of $\mathrm{VN}(\mathbb{H}^d)$ defined by (1.3.7) maps $C^*(\{R_k\}_{k=1}^{2d})$ into itself.*

The proof of Theorem 3.1.1 is given in Sect. 3.1.1.
The following result is the main result of this section.

Theorem 3.1.2 *Let* $A : \Omega \to \mathrm{Aut}(\mathbb{H}^d)$ *be a continuous function defined on an open subset* $\Omega \subset \mathbb{H}^d$. *The $*$-automorphism π_A given in Definition 1.3.16 maps $C(\Omega) \otimes_{\min} C^*(\{R_k\}_{k=1}^{2d})$ into itself.*

The proof of Theorem 3.1.2 is given in Sect. 3.1.4. Before, in Sects. 3.1.2 and 3.1.3, we prove some technical results, which we need.

3.1.1 Proof of Theorem 3.1.1

Notation 3.1.3 Let $A = (a_{ij})_{i,j=1}^{2d} \in \text{GL}(2d, \mathbb{R})$. We denote

$$X_i^A = \sum_{j=1}^{2d} a_{ij} X_j, \quad 1 \le i \le 2d, \quad P^A = \sum_{i=1}^{2d} (X_i^A)^2,$$

$$R_i^A = X_i^A (P^A)^{-\frac{1}{2}}, \quad 1 \le i \le 2d.$$

By Lemma 2.1.7, it is easy to see that if $A = (a_{ij})_{i,j=1}^{2d+1} \in \text{Aut}(\mathbb{H}^d)$, then

$$R_i^A = \pi_A(R_i), \quad 1 \le i \le 2d.$$

(Recall that we identify $A \in \text{Aut}(\mathbb{H}^d)$ with the corresponding matrix $S \in \text{GL}(2d, \mathbb{R})$ given by Theorem 2.1.6, denoting it by the same letter A.) To prove Theorem 3.1.1, it suffices to show that $R_i^A \in C^*(\{R_k\}_{k=1}^{2d})$. First, we establish two technical results.

Lemma 3.1.4 *Let L_1, L_2 be positive self-adjoint boundedly invertible operators on a Hilbert space H such that $\text{dom}(L_1^s) = \text{dom}(L_2^s)$ for every $s \ge 0$. If the operator $L_1^{-1}[L_1, L_2]$ is bounded, then so is*[1] $[L_1^{-\frac{1}{4}}, L_2^{\frac{1}{2}}]$.

Proof We use the following abstract equality (here $B(\cdot, \cdot)$ is the beta function)

$$x^{-\theta} = \frac{1}{B(\theta, 1-\theta)} \int_0^\infty \frac{\lambda^{-\theta} d\lambda}{x + \lambda}, \quad 0 < \theta < 1.$$

Thus,

$$L_1^{-\frac{1}{4}} = \frac{1}{B(\frac{1}{4}, \frac{3}{4})} \int_0^\infty \frac{1}{L_1 + \lambda_1} \lambda_1^{-\frac{1}{4}} d\lambda_1, \quad L_2^{\frac{1}{2}} = \frac{1}{B(\frac{1}{2}, \frac{1}{2})} \int_0^\infty \frac{L_2}{L_2 + \lambda_2} \lambda_2^{-\frac{1}{2}} d\lambda_2.$$

We, therefore, have

$$[L_1^{-\frac{1}{4}}, L_2^{\frac{1}{2}}] = \frac{1}{B(\frac{1}{4}, \frac{3}{4}) \cdot B(\frac{1}{2}, \frac{1}{2})} \int_0^\infty \int_0^\infty [\frac{L_2}{L_2 + \lambda_2}, \frac{1}{L_1 + \lambda_1}] \lambda_2^{-\frac{1}{2}} \lambda_1^{-\frac{1}{4}} d\lambda_1 d\lambda_2.$$

Now,

$$\frac{L_2}{L_2 + \lambda_2} = 1 - \frac{\lambda_2}{L_2 + \lambda_2}.$$

---

[1] More precisely, if $L_1^{-1}[L_1, L_2] : \cap_{s \ge 0} \text{dom}(L_1^s) \to H$ extends to a bounded operator on H, then so does $[L_1^{-\frac{1}{4}}, L_2^{\frac{1}{2}}] : \cap_{s \ge 0} \text{dom}(L_1^s) \to H$.

3.1 The Action on the Codomain of the Principal Symbol Map

Hence,

$$[\frac{L_2}{L_2+\lambda_2}, \frac{1}{L_1+\lambda_1}] = -\lambda_2 [\frac{1}{L_2+\lambda_2}, \frac{1}{L_1+\lambda_1}]$$

$$= \frac{\lambda_2}{L_2+\lambda_2} [L_2, \frac{1}{L_1+\lambda_1}] \frac{1}{L_2+\lambda_2}$$

$$= \frac{\lambda_2}{L_2+\lambda_2} \cdot \frac{1}{L_1+\lambda_1} [L_1, L_2] \cdot \frac{1}{L_1+\lambda_1} \frac{1}{L_2+\lambda_2}.$$

It follows that

$$\left\| [\frac{L_2}{L_2+\lambda_2}, \frac{1}{L_1+\lambda_1}] \right\|_\infty \le \left\| \frac{1}{L_1+\lambda_1} [L_1, L_2] \right\|_\infty \left\| \frac{1}{L_1+\lambda_1} \right\|_\infty \left\| \frac{1}{L_2+\lambda_2} \right\|_\infty.$$

Since $L_1, L_2 \ge c$ for some constant $c > 0$, it follows that

$$\left\| [\frac{L_2}{L_2+\lambda_2}, \frac{1}{L_1+\lambda_1}] \right\|_\infty \le \left\| L_1^{-1}[L_1, L_2] \right\|_\infty \cdot \frac{1}{c+\lambda_1} \cdot \frac{1}{c+\lambda_2}.$$

The assertion follows immediately. $\square$

Lemma 3.1.5 *Let $\mathcal{T}$ be a C^*-subalgebra in $B(H)$ containing $\mathcal{K}(H)$. Let L_1, L_2 be positive self-adjoint boundedly invertible operators on H such that $\mathrm{dom}(L_1^s) = \mathrm{dom}(L_2^s)$ for every $s \ge 0$. Suppose that*

(i) $c^{-1}L_1 \le L_2 \le cL_1$ *with some* $c > 0$;
(ii) $L_1^{-\frac12} L_2 L_1^{-\frac12} \in \mathcal{T}$;
(iii) $L_1^{-1}[L_1, L_2]$ *is bounded and* L_1^{-1} *is compact;*

We have $L_1^{\frac12} L_2^{-\frac12} \in \mathcal{T}$.

Proof We write

$$L_1^{-\frac14} L_2^{\frac12} L_1^{-\frac14} = L_2^{\frac12} L_1^{-\frac12} + [L_1^{-\frac14}, L_2^{\frac12}] \cdot L_1^{-\frac14}.$$

Using Lemma 3.1.4 and compactness of L_1^{-1}, we obtain

$$L_1^{-\frac14} L_2^{\frac12} L_1^{-\frac14} \in L_2^{\frac12} L_1^{-\frac12} + \mathcal{K}(H). \tag{3.1.1}$$

Taking adjoints, we obtain

$$L_1^{-\frac14} L_2^{\frac12} L_1^{-\frac14} \in L_1^{-\frac12} L_2^{\frac12} + \mathcal{K}(H).$$

Thus,
$$(L_1^{-\frac{1}{4}} L_2^{\frac{1}{2}} L_1^{-\frac{1}{4}})^2 \in L_1^{-\frac{1}{2}} L_2^{\frac{1}{2}} \cdot L_2^{\frac{1}{2}} L_1^{-\frac{1}{2}} + \mathcal{K}(H).$$

Since $\mathcal{K}(H) \subset \mathcal{T}$ and $L_1^{-\frac{1}{2}} L_2 L_1^{-\frac{1}{2}} \in \mathcal{T}$, it follows that

$$(L_1^{-\frac{1}{4}} L_2^{\frac{1}{2}} L_1^{-\frac{1}{4}})^2 \in \mathcal{T}.$$

Since $\mathcal{T}$ is a C^*-subalgebra and $L_1^{-\frac{1}{4}} L_2^{\frac{1}{2}} L_1^{-\frac{1}{4}} \geq 0$, it follows that

$$L_1^{-\frac{1}{4}} L_2^{\frac{1}{2}} L_1^{-\frac{1}{4}} \in \mathcal{T}. \tag{3.1.2}$$

Combining (3.1.1), (3.1.2) and taking into account that $\mathcal{K}(H) \subset \mathcal{T}$, we arrive at

$$L_2^{\frac{1}{2}} L_1^{-\frac{1}{2}} \in \mathcal{T}.$$

Since the operator on the left hand side is boundedly invertible and $\mathcal{T}$ is a C^*-algebra, the assertion follows. □

Proof of Theorem 3.1.1 As mentioned above, it suffices to show that $R_i^A \in C^*(\{R_k\}_{k=1}^{2d})$ for any $A = (a_{ij})_{i,j=1}^{2d+1} \in \mathrm{Aut}(\mathbb{H}^d)$.

First, we assume that A is diagonal. In the proof, we need the reduced Schrödinger representation π_{red}. □

Definition 3.1.6 The C^*-algebra $\mathcal{T}_d$ is the C^*-subalgebra in $B(L_2(\mathbb{R}^d)) \otimes \mathbb{C}^2$ generated by the operators $\{p_k H_d^{-\frac{1}{2}} \otimes (1,1)\}_{k=1}^d$ and $\{q_k H_d^{-\frac{1}{2}} \otimes (-1,1)\}_{k=1}^d$. Here, $H_d = \sum_{j=1}^d p_j^2 + q_j^2$.

We have (cf. (2.1.4))

$$C^*(\{R_k\}_{k=1}^{2d}) = \pi_{\mathrm{red}}(\mathcal{T}_d). \tag{3.1.3}$$

We claim that the operators

$$L_1 = H_d = \sum_{j=1}^d p_j^2 + q_j^2, \quad L_2 = P^A = \sum_{j=1}^d a_{j,j}^2 p_j^2 + a_{j+d,j+d}^2 q_j^2,$$

and the C^*-algebra

$$\mathcal{T} = \{X \in B(L_2(\mathbb{R}^d)) : X \otimes (1,1) \in \mathcal{T}_d\}$$

satisfy the assumptions in Lemma 3.1.5.

3.1 The Action on the Codomain of the Principal Symbol Map

It is clear that L_1 and L_2 are positive self-adjoint boundedly invertible operators, and L_1^{-1} is compact. Let us verify the condition on domains. By taking a scalar multiple of *diagonal* $A \in \mathrm{Aut}(\mathbb{H}^d)$ if needed, we may assume without loss of generality that $a_{j+d,j+d} = a_{j,j}^{-1}$ for every $1 \leq j \leq d$. If $U : L_2(\mathbb{R}^d) \to L_2(\mathbb{R}^d)$ is the anisotropic dilation by the factor $a_{j,j}^{\frac{1}{2}}$ in the j-th direction, then

$$U^{-1} p_j U = a_{j,j}^{\frac{1}{2}} p_j, \quad U^{-1} q_j U = a_{j,j}^{-\frac{1}{2}} q_j.$$

Thus, $U^{-1} H_d U = L_2$. This means

$$\mathrm{dom}(L_2^m) = \{\xi \in L_2(\mathbb{R}^d) : U\xi \in \mathrm{dom}(H_d^m)\}.$$

However, it is well known (see, for instance, [37, Proposition 1.6.6]) that

$$\mathrm{dom}(H_d^m) = \{\xi \in L_2(\mathbb{R}^d) : (p_1^{k_1} \cdots p_d^{k_d} q_1^{l_1} \cdots q_d^{l_d})\xi \in L_2(\mathbb{R}^d), \sum_{i=1}^d k_i + l_i \leq 2m\}.$$

In particular, U and U^{-1} map $\mathrm{dom}(H_d^m)$ into itself. Thus, $\mathrm{dom}(L_2^m) = \mathrm{dom}(L_1^m)$ for every $m \in \mathbb{Z}_+$. By complex interpolation, $\mathrm{dom}(L_2^s) = \mathrm{dom}(L_1^s)$ for every $s \geq 0$.

It follows from Lemma 5.2 in [29] that $\mathcal{K}(L_2(\mathbb{R}^d)) \subset \mathcal{T}$. The condition (i) in Lemma 3.1.5 is obvious. The condition (ii) in Lemma 3.1.5 can be seen as follows:

$$L_1^{-\frac{1}{2}} L_2 L_1^{-\frac{1}{2}} = \sum_{j=1}^d a_{j,j}^2 H_d^{-\frac{1}{2}} p_j^2 H_d^{-\frac{1}{2}} + \sum_{j=1}^d a_{j+d,j+d}^2 H_d^{-\frac{1}{2}} q_j^2 H_d^{-\frac{1}{2}}.$$

Clearly,

$$H_d^{-\frac{1}{2}} p_j^2 H_d^{-\frac{1}{2}} \otimes (1,1) = (p_j H_d^{-\frac{1}{2}} \otimes (1,1))^* \cdot (p_j H_d^{-\frac{1}{2}} \otimes (1,1)) \in \mathcal{T}_d,$$

$$H_d^{-\frac{1}{2}} q_j^2 H_d^{-\frac{1}{2}} \otimes (1,1) = (q_j H_d^{-\frac{1}{2}} \otimes (-1,1))^* \cdot (q_j H_d^{-\frac{1}{2}} \otimes (-1,1)) \in \mathcal{T}_d.$$

Thus,

$$H_d^{-\frac{1}{2}} p_j^2 H_d^{-\frac{1}{2}}, H_d^{-\frac{1}{2}} q_j^2 H_d^{-\frac{1}{2}} \in \mathcal{T}.$$

The condition (iii) in Lemma 3.1.5 can be seen as follows:

$$L_1^{-1}[L_1, L_2] = H_d^{-1} \sum_{j=1}^d a_{j,j}^2 [q_j^2, p_j^2] + a_{j+d,j+d}^2 [p_j^2, q_j^2]$$

$$= 2\iota H_d^{-1} \sum_{j=1}^d (p_j q_j + q_j p_j) \cdot (a_{j,j}^2 - a_{j+d,j+d}^2) \in B(L_2(\mathbb{R}^d)).$$

By Lemma 3.1.5, we have
$$L_1^{\frac{1}{2}} L_2^{-\frac{1}{2}} \otimes (1,1) \in \mathcal{T}_d.$$

Thus,
$$\Delta^{\frac{1}{2}}(P^A)^{-\frac{1}{2}} = \pi_{\text{red}}(L_1^{\frac{1}{2}} L_2^{-\frac{1}{2}} \otimes (1,1)) \in \pi_{\text{red}}(\mathcal{T}_d) = C^*(\{R_k\}_{k=1}^{2d}).$$

Consequently,
$$\pi_A(R_i) = R_i^A = X_i^A \Delta^{-\frac{1}{2}} \cdot \Delta^{\frac{1}{2}}(P^A)^{-\frac{1}{2}} \in C^*(\{R_k\}_{k=1}^{2d}), \quad 1 \le i \le 2d.$$

This immediately yields the assertion in the case when A is diagonal.

Now consider the case when $A \in \text{Aut}(\mathbb{H}^d)$ is arbitrary. Let S and λ be as in Theorem 2.1.6. If $\lambda > 0$, then $S_0 = \lambda^{-\frac{1}{2}} S \in \text{Sp}(2d, \mathbb{R})$ and, therefore,

$$A = \begin{pmatrix} S_0 & 0 \\ 0 & 1 \end{pmatrix} \cdot \begin{pmatrix} \lambda^{\frac{1}{2}} 1_{2d} & 0 \\ 0 & \lambda \end{pmatrix}.$$

If $\lambda < 0$, then
$$S_0 = |\lambda|^{-\frac{1}{2}} S \cdot \begin{pmatrix} 1_d & 0 \\ 0 & -1_d \end{pmatrix} \in \text{Sp}(2d, \mathbb{R}).$$

Thus,
$$A = \begin{pmatrix} S_0 & 0 \\ 0 & 1 \end{pmatrix} \cdot \begin{pmatrix} |\lambda|^{\frac{1}{2}} 1_{2d} & 0 \\ 0 & |\lambda| \end{pmatrix} \cdot \begin{pmatrix} 1_d & 0 & 0 \\ 0 & -1_d & 0 \\ 0 & 0 & -1 \end{pmatrix}.$$

By the composition rule (1.3.8), it suffices to prove the assertion for
$$A = \begin{pmatrix} S & 0 \\ 0 & 1 \end{pmatrix}, \quad S \in \text{Sp}(2d, \mathbb{R}).$$

Recall that $U(d) = \text{SO}(2d, \mathbb{R}) \cap \text{Sp}(2d, \mathbb{R})$. By Euler decomposition (see e.g. Propositions 32–34 in [20]), every $S \in \text{Sp}(2d, \mathbb{R})$ can be written as $S = U_1 D U_2$, where $U_1, U_2 \in U(d)$ and $D \in \text{Sp}(2d, \mathbb{R})$ is diagonal. By the composition rule (1.3.8), it suffices to prove the assertion

for $A = \begin{pmatrix} U & 0 \\ 0 & 1 \end{pmatrix}$, $U \in U(d)$, and for $A = \begin{pmatrix} D & 0 \\ 0 & 1 \end{pmatrix}$, diagonal $D \in \text{Sp}(2d, \mathbb{R})$.

3.1 The Action on the Codomain of the Principal Symbol Map

In the first case, $P^A = \Delta$ and the assertion is obvious. In the second case, the assertion is established in the first part of the proof.

3.1.2 Square Root Lemmas

Theorem 3.1.7 *Let P and Q be positive self-adjoint operators with the same domains and with trivial kernels on a Hilbert space H. We have*

$$\|(Q^{\frac{1}{2}} - P^{\frac{1}{2}})P^{-\frac{1}{2}}\|_\infty \leq c_{abs}\|Q^{-\frac{1}{4}}(Q - P)P^{-\frac{3}{4}}\|_\infty.$$

The rest of the subsection is devoted to the proof of this theorem.

Lemma 3.1.8 *Let x be a positive, self-adjoint operator with trivial kernel on H. For every $\xi \in \mathrm{dom}(x^{\frac{1}{2}})$, we have*

$$x^{\frac{1}{2}}\xi = \lim_{n\to\infty} \frac{2}{\pi}\int_0^n \frac{x}{x + \lambda^2}\xi\, d\lambda,$$

where the limit is taken in the norm of H.

Proof We have

$$\frac{2}{\pi}\int_0^n \frac{x^{\frac{1}{2}}}{x + \lambda^2}d\lambda = \frac{2}{\pi}\tan^{-1}(nx^{-\frac{1}{2}}).$$

Clearly, there exists the limit (in strong operator topology)

$$\lim_{n\to\infty} \frac{2}{\pi}\tan^{-1}(nx^{-\frac{1}{2}}) = 1.$$

Thus,

$$\lim_{n\to\infty} \frac{2}{\pi}\int_0^n \frac{x^{\frac{1}{2}}}{x + \lambda^2}\eta\, d\lambda = \eta$$

for every $\eta \in H$ (the limit is taken in the norm of H). Setting $\eta = x^{\frac{1}{2}}\xi$, we complete the proof. □

In this section, the notion of double operator integral (denoted by $T_F^{A,B}$) plays a crucial role. We refer the reader to [55] for the definition and basic properties of double operator integrals.

Lemma 3.1.9 *Let P and Q be positive self-adjoint operators bounded from below by strictly positive constants. For every $V \in B(H)$ and $n \in \mathbb{N}$, we have*

$$\frac{2}{\pi}\int_0^n \frac{Q^{\frac{1}{4}}}{Q+\lambda^2} V \frac{P^{\frac{1}{4}}}{P+\lambda^2} \lambda^2 d\lambda = T_\Theta^{\frac{1}{n}Q^{\frac{1}{2}}, \frac{1}{n}P^{\frac{1}{2}}}\left(T_{f^{[1]}}^{\frac{1}{n}Q^{\frac{1}{2}}, \frac{1}{n}P^{\frac{1}{2}}}(V)\right).$$

Here,

$$\Theta(\alpha_0, \alpha_1) = \frac{\alpha_0^{\frac{1}{2}}\alpha_1^{\frac{1}{2}}}{\alpha_0+\alpha_1}, \quad f(\alpha) = \frac{2}{\pi}\alpha \cot^{-1}(\alpha), \quad \alpha, \alpha_0, \alpha_1 > 0,$$

and $f^{[1]}$ denotes the first divided difference of f.

Proof Denote, for brevity,

$$\Theta_n(\alpha_0, \alpha_1) = \int_0^n \frac{\alpha_0^{\frac{1}{2}}}{\alpha_0^2+\lambda^2}\cdot\frac{\alpha_1^{\frac{1}{2}}}{\alpha_1^2+\lambda^2}\lambda^2 d\lambda.$$

It is immediate that

$$\int_0^n \frac{Q^{\frac{1}{4}}}{Q+\lambda^2} V \frac{P^{\frac{1}{4}}}{P+\lambda^2}\lambda^2 d\lambda = T_{\Theta_n}^{Q^{\frac{1}{2}}, P^{\frac{1}{2}}}(V).$$

On the other hand, we have

$$\Theta_n(\alpha_0, \alpha_1) = \Theta_1(\frac{\alpha_0}{n}, \frac{\alpha_1}{n}).$$

Thus,

$$\int_0^n \frac{Q^{\frac{1}{4}}}{Q+\lambda^2} V \frac{P^{\frac{1}{4}}}{P+\lambda^2}\lambda^2 d\lambda = T_{\Theta_1}^{\frac{1}{n}Q^{\frac{1}{2}}, \frac{1}{n}P^{\frac{1}{2}}}(V).$$

We now compute Θ_1. Clearly,

$$\frac{\alpha_0^{\frac{1}{2}}}{\alpha_0^2+\lambda^2}\cdot\frac{\alpha_1^{\frac{1}{2}}}{\alpha_1^2+\lambda^2}\cdot\lambda^2 = \frac{\alpha_0^{\frac{1}{2}}\alpha_1^{\frac{1}{2}}}{\alpha_0^2-\alpha_1^2}\cdot\left(\frac{\alpha_0^2}{\alpha_0^2+\lambda^2}-\frac{\alpha_1^2}{\alpha_1^2+\lambda^2}\right).$$

Thus,

$$\Theta_1(\alpha_0, \alpha_1) = \frac{\alpha_0^{\frac{1}{2}}\alpha_1^{\frac{1}{2}}}{\alpha_0^2-\alpha_1^2}\cdot\left(\alpha_0\cot^{-1}(\alpha_0)-\alpha_1\cot^{-1}(\alpha_1)\right) = \Theta(\alpha_0, \alpha_1)\cdot f^{[1]}(\alpha_0, \alpha_1).$$

This completes the proof. □

3.1 The Action on the Codomain of the Principal Symbol Map

Lemma 3.1.10 *Let A and B be positive self-adjoint operators. Let Θ and f be as in Lemma 3.1.9. We have*

$$\|T_\Theta^{A,B}\|_{\mathcal{L}_\infty \to \mathcal{L}_\infty} \leq \frac{1}{2}, \quad \|T_{f^{[1]}}^{A,B}\|_{\mathcal{L}_\infty \to \mathcal{L}_\infty} \leq c_{\text{abs}}.$$

Proof Clearly,

$$\Theta(\alpha_0, \alpha_1) = F\left(\frac{\alpha_0}{\alpha_1}\right),$$

where

$$F(e^\alpha) = \frac{e^{\frac{\alpha}{2}}}{e^\alpha + 1} = \frac{1}{2\cosh(\frac{\alpha}{2})}, \quad \alpha \in \mathbb{R}.$$

Hence, $F \circ \exp$ is a Schwartz function. It follows from Lemma 9 in [55] that

$$\|T_\Theta^{A,B}\|_{\mathcal{L}_\infty \to \mathcal{L}_\infty} \leq \frac{1}{\sqrt{2\pi}} \|\widehat{F \circ \exp}\|_1.$$

However, $\widehat{F \circ \exp}$ happens to be positive. Thus,

$$\frac{1}{\sqrt{2\pi}} \|\widehat{F \circ \exp}\|_1 = (F \circ \exp)(0) = \frac{1}{2}.$$

To prove the second assertion, we extend the function f to the real line such that f is everywhere smooth and supported on some semi-axis. It is immediate that $f' \in L_2(\mathbb{R})$ and $f'' \in L_2(\mathbb{R})$. By Lemma 7 in [55], the Fourier transform $\widehat{f'}$ of f' belongs to $L_1(\mathbb{R})$. Thus,

$$\|T_{f^{[1]}}^{A,B}\|_{\mathcal{L}_\infty \to \mathcal{L}_\infty} \leq \|\widehat{f'}\|_{L_1(\mathbb{R})} < \infty.$$

This completes the proof. □

Lemma 3.1.11 *Let P and Q be positive self-adjoint operators with the same domains. Suppose P and Q are bounded from below by strictly positive constants. We have*

$$\|(Q^{\frac{1}{2}} - P^{\frac{1}{2}})P^{-\frac{1}{2}}\|_\infty \leq c_{\text{abs}} \|Q^{-\frac{1}{4}}(Q - P)P^{-\frac{3}{4}}\|_\infty.$$

Proof Let $\xi \in H$ and note that $P^{-\frac{1}{2}}\xi \in \text{dom}(P^{\frac{1}{2}})$. By Lemma 3.1.8, we have

$$P^{\frac{1}{2}}(P^{-\frac{1}{2}}\xi) = \lim_{n\to\infty} \frac{2}{\pi} \int_0^n \frac{P}{P + \lambda^2} P^{-\frac{1}{2}}\xi \, d\lambda.$$

However, $\mathrm{dom}(P^{\frac{1}{2}}) = \mathrm{dom}(Q^{\frac{1}{2}})$. Hence, we also have $P^{-\frac{1}{2}}\xi \in \mathrm{dom}(Q^{\frac{1}{2}})$. By Lemma 3.1.8, we have

$$Q^{\frac{1}{2}}(P^{-\frac{1}{2}}\xi) = \lim_{n\to\infty} \frac{2}{\pi} \int_0^n \frac{Q}{Q+\lambda^2} P^{-\frac{1}{2}}\xi \, d\lambda.$$

Thus,

$$(Q^{\frac{1}{2}} - P^{\frac{1}{2}})P^{-\frac{1}{2}} = \lim_{n\to\infty} \frac{2}{\pi} \int_0^n \left(\frac{Q}{Q+\lambda^2} - \frac{P}{P+\lambda^2}\right) P^{-\frac{1}{2}} d\lambda,$$

where the limit is taken in strong operator topology.

We write

$$\left(\frac{Q}{Q+\lambda^2} - \frac{P}{P+\lambda^2}\right) P^{-\frac{1}{2}} = -\left(\frac{1}{Q+\lambda^2} - \frac{1}{P+\lambda^2}\right) \cdot \lambda^2 P^{-\frac{1}{2}}$$

$$= \frac{1}{Q+\lambda^2}(Q-P) \frac{\lambda^2 P^{-\frac{1}{2}}}{P+\lambda^2}$$

$$= \frac{Q^{\frac{1}{4}}}{Q+\lambda^2} \cdot Q^{-\frac{1}{4}}(Q-P)P^{-\frac{3}{4}} \cdot \frac{\lambda^2 P^{\frac{1}{4}}}{P+\lambda^2}.$$

Denote for brevity

$$V = Q^{-\frac{1}{4}}(Q-P)P^{-\frac{3}{4}}.$$

We have

$$\left(\frac{Q}{Q+\lambda^2} - \frac{P}{P+\lambda^2}\right) P^{-\frac{1}{2}} = \frac{Q^{\frac{1}{4}}}{Q+\lambda^2} V \frac{P^{\frac{1}{4}}}{P+\lambda^2} \lambda^2.$$

Hence,

$$\left(Q^{\frac{1}{2}} - P^{\frac{1}{2}}\right) P^{-\frac{1}{2}} = \lim_{n\to\infty} \frac{2}{\pi} \int_0^n \frac{Q^{\frac{1}{4}}}{Q+\lambda^2} V \frac{P^{\frac{1}{4}}}{P+\lambda^2} \lambda^2 d\lambda,$$

where the limit is taken in strong operator topology. By Lemma 3.1.9, we have (see Lemma 3.1.9 for the notations Θ and f)

$$\left(Q^{\frac{1}{2}} - P^{\frac{1}{2}}\right) P^{-\frac{1}{2}} = \lim_{n\to\infty} T_\Theta^{\frac{1}{n}Q^{\frac{1}{2}}, \frac{1}{n}P^{\frac{1}{2}}} \left(T_{f^{[1]}}^{\frac{1}{n}Q^{\frac{1}{2}}, \frac{1}{n}P^{\frac{1}{2}}}(V)\right),$$

3.1 The Action on the Codomain of the Principal Symbol Map

where the limit is taken in strong operator topology. By Lemma 3.1.10, we have

$$\left\| T_\Theta^{\frac{1}{n}Q^{\frac{1}{2}},\frac{1}{n}P^{\frac{1}{2}}} \left(T_{f[1]}^{\frac{1}{n}Q^{\frac{1}{2}},\frac{1}{n}P^{\frac{1}{2}}}(V) \right) \right\|_\infty \leq c_{\text{abs}} \|V\|_\infty.$$

It follows now from the Fatou property in $\mathcal{L}_\infty$ that

$$\left\| \left(Q^{\frac{1}{2}} - P^{\frac{1}{2}} \right) P^{-\frac{1}{2}} \right\|_\infty \leq c_{\text{abs}} \|V\|_\infty.$$

This completes the proof. □

Proof of Theorem 3.1.7 Applying Lemma 3.1.11, we obtain

$$\|((Q+\epsilon)^{\frac{1}{2}} - (P+\epsilon)^{\frac{1}{2}})(P+\epsilon)^{-\frac{1}{2}}\|_\infty \leq c_{\text{abs}} \|(Q+\epsilon)^{-\frac{1}{4}}(Q-P)(P+\epsilon)^{-\frac{3}{4}}\|_\infty.$$

Clearly,

$$\|(Q+\epsilon)^{-\frac{1}{4}}(Q-P)(P+\epsilon)^{-\frac{3}{4}}\|_\infty$$
$$\leq \|(Q+\epsilon)^{-\frac{1}{4}} Q^{\frac{1}{4}}\|_\infty \|Q^{-\frac{1}{4}}(Q-P)P^{-\frac{3}{4}}\|_\infty \|P^{\frac{1}{4}}(P+\epsilon)^{-\frac{1}{4}}\|_\infty.$$

Thus,

$$\|((Q+\epsilon)^{\frac{1}{2}} - (P+\epsilon)^{\frac{1}{2}})(P+\epsilon)^{-\frac{1}{2}}\|_\infty \leq c_{\text{abs}} \|Q^{-\frac{1}{4}}(Q-P)P^{-\frac{3}{4}}\|_\infty.$$

Set

$$A_\epsilon = \epsilon^{-\frac{1}{2}} \cdot ((Q+\epsilon)^{\frac{1}{2}} - Q^{\frac{1}{2}}), \quad B_\epsilon = \epsilon^{\frac{1}{2}}(P+\epsilon)^{-\frac{1}{2}}, \quad C_\epsilon = P^{\frac{1}{2}}(P+\epsilon)^{-\frac{1}{2}}.$$

Note that $A_\epsilon, B_\epsilon \to 0$ and $C_\epsilon \to 1$ in strong operator topology. We have

$$(Q+\epsilon)^{\frac{1}{2}}(P+\epsilon)^{-\frac{1}{2}} = A_\epsilon B_\epsilon + Q^{\frac{1}{2}} P^{-\frac{1}{2}} C_\epsilon.$$

It follows that

$$(Q+\epsilon)^{\frac{1}{2}}(P+\epsilon)^{-\frac{1}{2}} \to Q^{\frac{1}{2}} P^{-\frac{1}{2}}$$

in strong operator topology. In other words,

$$((Q+\epsilon)^{\frac{1}{2}} - (P+\epsilon)^{\frac{1}{2}})(P+\epsilon)^{-\frac{1}{2}} \to (Q^{\frac{1}{2}} - P^{\frac{1}{2}})P^{-\frac{1}{2}}$$

in strong operator topology. The assertion follows now from the Fatou property in $\mathcal{L}_\infty$. □

3.1.3 Continuity Theorem

In this subsection, we prove the following continuity result, which allows us to complete the proof of Theorem 3.1.2.

Theorem 3.1.12 *Let $A, B \in \mathrm{GL}(2d, \mathbb{R})$. We have*

$$\|R_i^A - R_i^B\|_\infty \leq c_d \|A - B\|_\infty \cdot \max\{\|A\|_\infty, \|B\|_\infty\}^3 \cdot \max\{\|A^{-1}\|_\infty, \|B^{-1}\|_\infty\}^4.$$

The rest of the subsection is devoted to the proof of the above theorem.

Lemma 3.1.13 *If $A \in \mathrm{GL}(2d, \mathbb{R})$, then the operator $\Delta^{\frac{1}{2}}(P^A)^{-\frac{1}{2}}$ extends to a bounded operator in $L_2(\mathbb{H}^d)$ with the following norm estimate:*

$$\|\Delta^{\frac{1}{2}}(P^A)^{-\frac{1}{2}}\|_\infty \leq \|A^{-1}\|_\infty.$$

Proof For every $\xi \in W^{2,2}(\mathbb{H}^d)$, we have

$$\langle P^A \xi, \xi \rangle = \sum_{i=1}^{2d} \|\sum_{j=1}^{2d} a_{ij} \cdot X_j \xi\|_{L_2(\mathbb{H}^d)}^2 = \int_{\mathbb{H}^d} \|\{\sum_{j=1}^{2d} a_{ij}(X_j\xi)(p)\}_{i=1}^{2d}\|_{\mathbb{C}^{2d}}^2 dp.$$

Denote for brevity

$$z(p) = \{(X_j\xi)(p)\}_{j=1}^{2d}, \quad p \in \mathbb{H}^d.$$

This allows us to write

$$\langle P^A \xi, \xi \rangle = \int_{\mathbb{H}^d} \|Az(p)\|_{\mathbb{C}^{2d}}^2 dp.$$

It is immediate that

$$\|z(p)\|_{\mathbb{C}^{2d}} \leq \|A^{-1}\|_\infty \|Az(p)\|_{\mathbb{C}^{2d}}.$$

Thus,

$$\langle P^A \xi, \xi \rangle \geq \|A^{-1}\|_\infty^{-2} \cdot \int_{\mathbb{H}^d} \|z(p)\|_{\mathbb{C}^{2d}}^2 dp = \|A^{-1}\|_\infty^{-2} \cdot \langle \Delta \xi, \xi \rangle.$$

In other words,

$$\|(P^A)^{\frac{1}{2}}\xi\|^2 \geq \|A^{-1}\|_\infty^{-2} \|\Delta^{\frac{1}{2}}\xi\|^2, \quad \xi \in W^{2,2}(\mathbb{H}^d).$$

3.1 The Action on the Codomain of the Principal Symbol Map

The assertion follows by applying Fact 2.5.8 with $B_1 = \Delta^{\frac{1}{2}}$ and $B_2 = (P^A)^{\frac{1}{2}}$. This completes the proof. □

Lemma 3.1.14 *If $A \in \mathrm{GL}(2d, \mathbb{R})$, then $\|R_i^A\|_\infty = 1$.*

Proof It follows from Lemma 2.1.7 that $R_i^A = V_A^{-1} R_i V_A$. Since V_A is a scalar multiple of a unitary operator, the assertion follows. □

Lemma 3.1.15 *If $A \in \mathrm{GL}(2d, \mathbb{R})$, then*

$$\|\Delta^{\frac{1}{4}}(P^A)^{-\frac{1}{4}}\|_\infty \leq \|A^{-1}\|_\infty^{\frac{1}{2}}.$$

Proof The assertion follows from Lemma 3.1.13 and the Hadamard 3 lines lemma. □

Notation 3.1.16 If $B \in \mathrm{GL}(2d, \mathbb{R})$, then we denote

$$p_i^B = \sum_{j=1}^{d} b_{ij} p_j, \quad q_i^B = \sum_{j=d+1}^{2d} b_{ij} q_{j-d}, \quad H^B = \sum_{i=1}^{2d} |p_i^B + q_i^B|^2.$$

If $B = 1_{2d}$, then $p_i^B = p_i$ and $q_i^B = 0$ for $1 \leq i \leq d$ and $p_i^B = 0$ and $q_i^B = q_{i-d}$ for $d+1 \leq i \leq 2d$. Thus, $H^B = \sum_{i=1}^{d} p_i^2 + q_i^2 = H_d$.

Lemma 3.1.17 *Let $B \in \mathrm{GL}(2d, \mathbb{R})$. We have*

(i) $\|H_d^{\frac{1}{2}}(H^B)^{-\frac{1}{2}}\|_\infty \leq \|B^{-1}\|_\infty$.
(ii) $\|H_d(H^B)^{-1}\|_\infty \leq c_d \|B^{-1}\|_\infty^4 \|B\|_\infty^2$.
(iii) $\|H_d^{\frac{3}{4}}(H^B)^{-\frac{3}{4}}\|_\infty \leq c_d \|B\|_\infty \|B^{-1}\|_\infty^{\frac{5}{2}}$.

Proof If $\xi \in L_2(\mathbb{R}^d)$ is a Schwartz function, then

$$\langle H^B \xi, \xi \rangle = \sum_{i=1}^{2d} \left\|\left(\sum_{j=1}^{d} b_{ij} p_j + \sum_{j=d+1}^{2d} b_{ij} q_{j-d}\right)\xi\right\|^2$$

$$= \int_{\mathbb{R}^d} \|\{\sum_{j=1}^{d} b_{ij}(p_j \xi)(p) + \sum_{j=d+1}^{2d} b_{ij}(q_{j-d}\xi)(p)\}_{i=1}^{2d}\|_{\mathbb{C}^{2d}}^2 dp.$$

Set

$$z(p) = \{(p_1 \xi)(p), \cdots, (p_d \xi)(p), (q_1 \xi)(p), \cdots, (q_d \xi)(p)\}.$$

We have

$$\langle H^B \xi, \xi \rangle = \int_{\mathbb{R}^d} \|Bz(p)\|_{\mathbb{C}^{2d}}^2 dp.$$

Clearly,
$$\|Bz(p)\|_{\mathbb{C}^{2d}}^2 \geq \|B^{-1}\|_\infty^{-2} \|z(p)\|_{\mathbb{C}^{2d}}^2.$$

Hence,
$$\langle H^B \xi, \xi \rangle \geq \|B^{-1}\|_\infty^{-2} \int_{\mathbb{R}^d} \|z(p)\|_{\mathbb{C}^{2d}}^2 dp = \|B^{-1}\|_\infty^{-2} \langle H_d \xi, \xi \rangle.$$

In other words,
$$H^B \geq \|B^{-1}\|_\infty^{-2} H_d \text{ and } (H^B)^{-\frac{1}{2}} H_d (H^B)^{-\frac{1}{2}} \leq \|B^{-1}\|_\infty^2.$$

This proves the first assertion.

We have
$$H_d(H^B)^{-1} = \sum_{i=1}^d p_i^2 (H^B)^{-1} + q_i^2 (H^B)^{-1}$$
$$= \sum_{i=1}^d p_i (H^B)^{-1} p_i + q_i (H^B)^{-1} q_i$$
$$+ \sum_{i=1}^d p_i (H^B)^{-1} [H^B, p_i] (H^B)^{-1} + q_i (H^B)^{-1} [H^B, q_i] (H^B)^{-1}.$$

Clearly,
$$\|p_i(H^B)^{-1} p_i\|_\infty = \|p_i(H^B)^{-\frac{1}{2}}\|_\infty^2 \leq \|H_d^{\frac{1}{2}}(H^B)^{-\frac{1}{2}}\|_\infty^2 \leq \|B^{-1}\|_\infty^2,$$
$$\|q_i(H^B)^{-1} q_i\|_\infty = \|q_i(H^B)^{-\frac{1}{2}}\|_\infty^2 \leq \|H_d^{\frac{1}{2}}(H^B)^{-\frac{1}{2}}\|_\infty^2 \leq \|B^{-1}\|_\infty^2.$$

Next, we write
$$p_i(H^B)^{-1} [H^B, p_i] (H^B)^{-1} = p_i(H^B)^{-\frac{1}{2}} \cdot (H^B)^{-\frac{1}{2}} H_d^{\frac{1}{2}} \cdot H_d^{-\frac{1}{2}} [H^B, p_i] \cdot (H^B)^{-1},$$

where each factor is bounded. Thus,
$$\|p_i(H^B)^{-1} [H^B, p_i] (H^B)^{-1}\|_\infty$$
$$\leq \|p_i(H^B)^{-\frac{1}{2}}\|_\infty \|(H^B)^{-\frac{1}{2}} H_d^{\frac{1}{2}}\|_\infty \|H_d^{-\frac{1}{2}} [H^B, p_i]\|_\infty \|(H^B)^{-1}\|_\infty$$
$$\leq \|B^{-1}\|_\infty \cdot \|B^{-1}\|_\infty \cdot c'_{d,i} \|B\|_\infty^2 \cdot \|B^{-1}\|_\infty^2$$

3.1 The Action on the Codomain of the Principal Symbol Map

and

$$\|q_i(H^B)^{-1}[H^B, q_i](H^B)^{-1}\|_\infty$$
$$\leq \|q_i(H^B)^{-\frac{1}{2}}\|_\infty \|(H^B)^{-\frac{1}{2}} H_d^{\frac{1}{2}}\|_\infty \|H_d^{-\frac{1}{2}}[H^B, q_i]\|_\infty \|(H^B)^{-1}\|_\infty$$
$$\leq \|B^{-1}\|_\infty \cdot \|B^{-1}\|_\infty \cdot c''_{d,i} \|B\|_\infty^2 \cdot \|B^{-1}\|_\infty^2.$$

Thus,

$$\|H^{\mathrm{id}}(H^B)^{-1}\|_\infty \leq 2d\|B^{-1}\|_\infty^2 + \|B^{-1}\|_\infty^4 \|B\|_\infty^2 \cdot \sum_{i=1}^d (c'_{d,i} + c''_{d,i}).$$

This proves the second assertion.

The last assertion follows from the preceding two by the Hadamard 3 lines lemma. □

Lemma 3.1.18 *Let* $B \in \mathrm{GL}(2d, \mathbb{R})$. *We have*

$$\|\Delta^{\frac{3}{4}} (P^B)^{-\frac{3}{4}}\|_\infty \leq c_d \|B\|_\infty \|B^{-1}\|_\infty^{\frac{5}{2}}.$$

Proof By Corollary 2.1.5, we have

$$\Delta = \pi(H_d \otimes |s|),$$

and, by Proposition 2.1.4,

$$X_j = \pi(\iota p_j \otimes |s|^{\frac{1}{2}}), \quad X_{j+d} = \pi(\iota q_j \otimes \mathrm{sgn}(s)|s|^{\frac{1}{2}}), \quad 1 \leq j \leq d.$$

Denote for brevity $V = \sum_{j=1}^d E_{j,j} - E_{j+d,j+d}$. We now write

$$X_i^B = \sum_{j=1}^{2d} b_{i,j} X_j$$
$$= \sum_{j=1}^d b_{i,j} X_j + b_{i,j+d} X_{j+d}$$
$$= \pi(\iota(p_i^B + q_i^B) \otimes |s|^{\frac{1}{2}} \chi_{(0,\infty)}(s)) + \pi(\iota(p_i^{BV} + q_i^{BV}) \otimes |s|^{\frac{1}{2}} \chi_{(-\infty,0)}(s))$$

and

$$P_B = \sum_{i=1}^{2d} |X_i^B|^2 = \pi(H^B \otimes |s|\chi_{(0,\infty)}(s) + H^{BV} \otimes |s|\chi_{(-\infty,0)}(s)).$$

Thus,
$$\Delta^{\frac{3}{4}}(P^B)^{-\frac{3}{4}} = \pi\big(H_d^{\frac{3}{4}}(H^B)^{-\frac{3}{4}} \otimes \chi_{(0,\infty)}(s) + H_d^{\frac{3}{4}}(H^{BV})^{-\frac{3}{4}} \otimes \chi_{(-\infty,0)}(s)\big).$$

Hence,
$$\|\Delta^{\frac{3}{4}}(P^B)^{-\frac{3}{4}}\|_\infty \leq \max\{\|H_d^{\frac{3}{4}}(H^B)^{-\frac{3}{4}}\|_\infty, \|H_d^{\frac{3}{4}}(H^{BV})^{-\frac{3}{4}}\|_\infty\}.$$

The assertion follows now from Lemma 3.1.17. □

Proof of Theorem 3.1.12 We write
$$R_i^A - R_i^B = (X_i^A - X_i^B)\Delta^{-\frac{1}{2}} \cdot \Delta^{\frac{1}{2}}(P^A)^{-\frac{1}{2}}$$
$$+ X_i^B \Delta^{-\frac{1}{2}} \cdot \Delta^{\frac{1}{2}}(P^A)^{-\frac{1}{2}} \cdot ((P^B)^{\frac{1}{2}} - (P^A)^{\frac{1}{2}})(P^B)^{-\frac{1}{2}}.$$

Taking norms, we obtain
$$\|R_i^A - R_i^B\|_\infty \leq \|(X_i^A - X_i^B)\Delta^{-\frac{1}{2}}\|_\infty \|\Delta^{\frac{1}{2}}(P^A)^{-\frac{1}{2}}\|_\infty$$
$$+ \|X_i^B \Delta^{-\frac{1}{2}}\|_\infty \|\Delta^{\frac{1}{2}}(P^A)^{-\frac{1}{2}}\|_\infty \|((P^B)^{\frac{1}{2}} - (P^A)^{\frac{1}{2}})(P^B)^{-\frac{1}{2}}\|_\infty.$$

Using Theorem 3.1.7 and Lemma 3.2.5, we write
$$\|R_i^A - R_i^B\|_\infty \leq \|A - B\|_\infty \|A^{-1}\|_{L_\infty(\mathbb{H}^d, M_{2d}(\mathbb{R}))}$$
$$+ c_{\text{abs}} \|B\|_\infty \|A^{-1}\|_\infty \|(P^A)^{-\frac{1}{4}}(P^A - P^B)(P^B)^{-\frac{3}{4}}\|_\infty. \quad (3.1.4)$$

Clearly,
$$(P^A)^{-\frac{1}{4}}(P^A - P^B)(P^B)^{-\frac{3}{4}} = (P^A)^{-\frac{1}{4}} \Delta^{\frac{1}{4}} \cdot \Delta^{-\frac{1}{4}}(P^A - P^B)\Delta^{-\frac{3}{4}} \cdot \Delta^{\frac{3}{4}}(P^B)^{-\frac{3}{4}}.$$

Taking norms, we obtain
$$\|(P^A)^{-\frac{1}{4}}(P^A - P^B)(P^B)^{-\frac{3}{4}}\|_\infty$$
$$\leq \|(P^A)^{-\frac{1}{4}}\Delta^{\frac{1}{4}}\|_\infty \|\Delta^{-\frac{1}{4}}(P^A - P^B)\Delta^{-\frac{3}{4}}\|_\infty \|\Delta^{\frac{3}{4}}(P^B)^{-\frac{3}{4}}\|_\infty.$$

The first factor is controlled by Lemma 3.1.15. It is immediate that
$$\|\Delta^{-\frac{1}{4}}(P^A - P^B)\Delta^{-\frac{3}{4}}\|_\infty \leq c_d(\|A\|_\infty + \|B\|_\infty)\|A - B\|_\infty.$$

The third factor is controlled by Lemma 3.1.18. Thus,

$$\|(P^A)^{-\frac{1}{4}}(P^A - P^B)(P^B)^{-\frac{3}{4}}\|_\infty$$

$$\leq c_d \|A^{-1}\|_\infty^{\frac{1}{2}} \|B\|_\infty \|B^{-1}\|_\infty^{\frac{5}{2}} \cdot (\|A\|_\infty + \|B\|_\infty) \cdot \|A - B\|_\infty. \qquad (3.1.5)$$

Combining the (3.1.4) and (3.1.5), we complete the proof. $\square$

3.1.4 Proof of Theorem 3.1.2

Let $A : \Omega \to \mathrm{Aut}(\mathbb{H}^d)$ be a continuous function defined on an open subset $\Omega \subset \mathbb{H}^d$. Clearly,

$$(\pi_A(1 \otimes R_i))(p) = R_i^{A(p)}, \quad p \in \Omega.$$

By Theorem 3.1.1, we have

$$R_i^{A(p)} \in C^*(\{R_k\}_{k=1}^{2d}), \quad p \in \Omega.$$

By Theorem 3.1.12, the mapping

$$p \in \Omega \mapsto R_i^{A(p)} \in C^*(\{R_k\}_{k=1}^{2d})$$

is continuous. It follows that

$$\pi_A(1 \otimes R_i) \in C(\Omega) \otimes_{\min} C^*(\{R_k\}_{k=1}^{2d}).$$

Since the C^*-algebra $C(\Omega) \otimes_{\min} C^*(\{R_k\}_{k=1}^{2d})$ is generated by the elements $1 \otimes R_i$, $1 \leq i \leq 2d$, and $M_\phi \otimes 1$, $\phi \in C(\Omega)$, Theorem 3.1.2 follows by continuity of π_A.

3.2 Main Computational Theorem

In this section, we work in a more general setting, which we will need later.

Notation 3.2.1 Let $A = (a_{ij})_{i,j=1}^{2d} : \mathbb{H}^d \to \mathrm{GL}(2d, \mathbb{R})$ be a smooth bounded mapping. We denote

$$X_i^A = \sum_{j=1}^{2d} M_{a_{ij}} X_j, \quad 1 \leq i \leq 2d, \quad P^A = \sum_{i=1}^{2d} |X_i^A|^2,$$

$$R_i^A = X_i^A (P^A)^{-\frac{1}{2}}, \quad 1 \leq i \leq 2d.$$

The main result is the following theorem.

Theorem 3.2.2 *Let $A : \mathbb{H}^d \to \mathrm{GL}(2d, \mathbb{R})$ be a smooth function, which is constant outside some ball. Let $\phi \in C_c^\infty(\mathbb{H}^d)$. We have*

$$M_\phi R_i^A M_\phi \in \Pi, \quad 1 \leq i \leq 2d,$$

$$\left(\mathrm{sym}(M_\phi R_i^A M_\phi)\right)(p) = \phi^2(p) R_i^{A(p)}, \quad p \in \mathbb{H}^d.$$

The proof of Theorem 3.2.2 is given in Sect. 3.2.4. Before, in Sects. 3.2.1, 3.2.2 and 3.2.3, we prove some technical results, which we need in the proof.

3.2.1 Continuity Theorem

First, we prove conditions on the function A, which ensure that the operators R_i^A are well defined.

Theorem 3.2.3 *Let $A : \mathbb{H}^d \to \mathrm{GL}(2d, \mathbb{R})$ be a smooth function, which is constant outside some ball. Then the operator R_i^A, $1 \leq i \leq 2d$, is a well-defined bounded operator in $L_2(\mathbb{H}^d)$ such that*

$$\|R_i^A\|_\infty \leq 2d \|A\|_{L_\infty(\mathbb{H}^d, M_{2d}(\mathbb{R}))} \|A^{-1}\|_{L_\infty(\mathbb{H}^d, M_{2d}(\mathbb{R}))}.$$

The main result of this subsection is the following theorem.

Theorem 3.2.4 *Let $A : \mathbb{H}^d \to \mathrm{GL}(2d, \mathbb{R})$ be a smooth function, which is constant outside some ball. For every $p \in \mathbb{H}^d$, we have*

$$\|R_i^A - R_i^{A(p)}\|_\infty \leq c_d (\|A - A(p)\|_{L_\infty(\mathbb{H}^d, M_{2d}(\mathbb{R}))} + \|A\|_{\dot{W}^{1,2d+2}(\mathbb{H}^d, M_{2d}(\mathbb{R}))})$$

$$\cdot \|A\|_{L_\infty(\mathbb{H}^d, M_{2d}(\mathbb{R}))}^3 \|A^{-1}\|_{L_\infty(\mathbb{H}^d, M_{2d}(\mathbb{R}))}^4.$$

The rest of this subsection is devoted to the proof of these theorems.

Let $A : \mathbb{H}^d \to \mathrm{GL}(2d, \mathbb{R})$ be a smooth function, which is constant outside some ball. The operator P^A is formally self-adjoint, formally elliptic differential operator of order 2 on $\mathbb{H}^d$. Its coefficients are real-valued and constant outside some ball. By Theorem 2.5.2, P^A is self-adjoint with domain $W^{2,2}(\mathbb{H}^d)$ and positive. By spectral theorem, the operator $(P^A)^{-\frac{1}{2}}$ is a well defined bounded operator in $L_2(\mathbb{H}^d)$.

Lemma 3.2.5 *Let $A : \mathbb{H}^d \to \mathrm{GL}(2d, \mathbb{R})$ be a smooth function, which is constant outside some ball. The operator $\Delta^{\frac{1}{2}}(P^A)^{-\frac{1}{2}}$ extends to a bounded operator in $L_2(\mathbb{H}^d)$ with the following norm estimate:*

$$\|\Delta^{\frac{1}{2}}(P^A)^{-\frac{1}{2}}\|_\infty \leq \|A^{-1}\|_{L_\infty(\mathbb{H}^d, M_{2d}(\mathbb{R}))}.$$

3.2 Main Computational Theorem

Proof For every $\xi \in W^{2,2}(\mathbb{H}^d)$, we have

$$\langle P^A \xi, \xi \rangle = \sum_{i=1}^{2d} \| \sum_{j=1}^{2d} a_{ij} \cdot X_j \xi \|^2_{L_2(\mathbb{H}^d)} = \int_{\mathbb{H}^d} \| \{ \sum_{j=1}^{2d} a_{ij}(p)(X_j\xi)(p) \}_{i=1}^{2d} \|^2_{\mathbb{C}^{2d}} dp.$$

Denote for brevity

$$z(p) = \{(X_j\xi)(p)\}_{j=1}^{2d}, \quad p \in \mathbb{H}^d.$$

This allows us to write

$$\langle P^A \xi, \xi \rangle = \int_{\mathbb{H}^d} \| A(p) z(p) \|^2_{\mathbb{C}^{2d}} dp.$$

It is immediate that

$$\|z(p)\|_{\mathbb{C}^{2d}} \leq \|A(p)^{-1}\|_\infty \|A(p)z(p)\|_{\mathbb{C}^{2d}} \leq \sup_{p \in \mathbb{H}^d} \|A^{-1}(p)\|_\infty \cdot \|A(p)z(p)\|_{\mathbb{C}^{2d}}.$$

Thus,

$$\langle P^A \xi, \xi \rangle \geq (\sup_{p \in \mathbb{H}^d} \|A^{-1}(p)\|_\infty)^{-2} \cdot \int_{\mathbb{H}^d} \|z(p)\|^2_{\mathbb{C}^{2d}} dp$$

$$= (\sup_{p \in \mathbb{H}^d} \|A^{-1}(p)\|_\infty)^{-2} \cdot \langle \Delta \xi, \xi \rangle.$$

In other words,

$$\|(P^A)^{\frac{1}{2}} \xi\|^2 \geq (\sup_{u \in \mathbb{H}^d} \|A^{-1}(p)\|_\infty)^{-2} \|\Delta^{\frac{1}{2}} \xi\|^2, \quad \xi \in W^{2,2}(\mathbb{H}^d).$$

The assertion follows by applying Fact 2.5.8 with $B_1 = \Delta^{\frac{1}{2}}$ and $B_2 = (P^A)^{\frac{1}{2}}$. This completes the proof. □

Proof of Theorem 3.2.3 We write

$$R_i^A = X_i^A \Delta^{-\frac{1}{2}} \cdot \Delta^{\frac{1}{2}} (P^A)^{-\frac{1}{2}} = \sum_{j=1}^{2d} M_{a_{ij}} R_j \cdot \Delta^{\frac{1}{2}} (P^A)^{-\frac{1}{2}}.$$

Taking norms, we obtain

$$\|R_i^A\|_\infty \leq \| \sum_{j=1}^{2d} M_{a_{ij}} R_j \|_\infty \| \Delta^{\frac{1}{2}} (P^A)^{-\frac{1}{2}} \|_\infty.$$

Using Lemma 3.2.5, we immediately complete the proof. □

Lemma 3.2.6 *Let $A : \mathbb{H}^d \to \mathrm{GL}(2d, \mathbb{R})$ be a smooth function, which is constant outside some ball. We have*

$$\|\Delta^{\frac14}(P^A)^{-\frac14}\|_\infty \leq \|A^{-1}\|_{L_\infty(\mathbb{H}^d, M_{2d}(\mathbb{R}))}^{\frac12}.$$

Proof The assertion follows from Lemma 3.2.5 and the Hadamard 3 lines lemma. □

Lemma 3.2.7 *Let $f : \mathbb{H}^d \to \mathbb{R}$ be a smooth bounded function. We have*

$$\|\Delta^{\frac14} M_f \Delta^{-\frac14}\|_\infty \leq c_d(\|f\|_\infty + \|f\|_{\dot{W}^{1,2d+2}}).$$

Proof By the Maximum Modulus Principle, we have

$$\|\Delta^{\frac14} M_f \Delta^{-\frac14}\|_\infty \leq \max\{\|\Delta^0 M_f \Delta^{-0}\|_\infty, \|\Delta^{\frac12} M_f \Delta^{-\frac12}\|_\infty\}.$$

By Hölder inequality, we have

$$\|\Delta^{\frac12} M_f \Delta^{-\frac12}\|_\infty \leq \|\Delta^{\frac12}(\sum_{i=1}^{2d}|X_i|)^{-1}\|_\infty \|(\sum_{i=1}^{2d}|X_i|) M_f \Delta^{-\frac12}\|_\infty$$

$$\leq c'_d \sum_{i=1}^{2d} \|X_i M_f \Delta^{-\frac12}\|_\infty.$$

Clearly,

$$X_i M_f \Delta^{-\frac12} = M_f R_i + M_{X_i f} \Delta^{-\frac12}.$$

Thus, taking into account that the uniform norm of an operator in $B(H)$ is controlled by its Schatten norm,

$$\|X_i M_f \Delta^{-\frac12}\|_\infty \leq \|M_f R_i\|_\infty + \|M_{X_i f} \Delta^{-\frac12}\|_\infty$$

$$\leq \|f\|_\infty + \|M_{X_i f} \Delta^{-\frac12}\|_{2d+2,\infty}.$$

The assertion follow now from Cwikel estimate (see Theorem 1.1 in [48]). □

Lemma 3.2.8 *Let $A = (a_{ij})_{i,j=1}^{2d}, B = (b_{ij})_{i,j=1}^{2d} : \mathbb{H}^d \to \mathrm{GL}(2d, \mathbb{R})$ be smooth bounded functions. We have*

$$\|\Delta^{-\frac14}(P^A - P^B)\Delta^{-\frac34}\|_\infty$$

$$\leq c_d(\|A\|_{L_\infty(\mathbb{H}^d, M_{2d}(\mathbb{R}))} + \|B\|_{L_\infty(\mathbb{H}^d, M_{2d}(\mathbb{R}))}) \cdot$$

$$\cdot (\|A - B\|_{L_\infty(\mathbb{H}^d, M_{2d}(\mathbb{R}))} + \|A\|_{\dot{W}^{1,2d+2}(\mathbb{H}^d, M_{2d}(\mathbb{R}))} + \|B\|_{\dot{W}^{1,2d+2}(\mathbb{H}^d, M_{2d}(\mathbb{R}))}).$$

3.2 Main Computational Theorem

Proof We write

$$P^A - P^B = \sum_{i=1}^{2d} \sum_{j_1,j_2=1}^{2d} X_{j_1} M_{b_{ij_1} b_{ij_2} - a_{ij_1} a_{ij_2}} X_{j_2}.$$

Hence,

$$\Delta^{-\frac{1}{4}}(P^A - P^B)\Delta^{-\frac{3}{4}} = \sum_{i=1}^{2d} \sum_{j_1,j_2=1}^{2d} \Delta^{-\frac{1}{4}} K_{j_1} T_{i,j_1,j_2} L_{j_2},$$

where

$$K_j = \Delta^{-\frac{1}{4}} X_j \Delta^{-\frac{1}{4}}, \quad L_j = \Delta^{\frac{1}{4}} X_j \Delta^{-\frac{3}{4}}, \quad 1 \leq j \leq 2d,$$

$$T_{i,j_1,j_2} = \Delta^{\frac{1}{4}} M_{b_{ij_1} b_{ij_2} - a_{ij_1} a_{ij_2}} \Delta^{-\frac{1}{4}}.$$

Thus,

$$\|\Delta^{-\frac{1}{4}}(P^A - P^B)\Delta^{-\frac{3}{4}}\|_\infty$$
$$\leq 8d^3 \sup_{1 \leq j \leq 2d} \|K_j\|_\infty \cdot \sup_{1 \leq j \leq 2d} \|L_j\|_\infty \cdot \sup_{1 \leq i,j_1,j_2 \leq 2d} \|T_{i,j_1,j_2}\|_\infty. \qquad (3.2.1)$$

It follows from Lemma 3.2.7 that

$$\|T_{i,j_1,j_2}\|_\infty \leq c'_d (\|b_{ij_1} b_{ij_2} - a_{ij_1} a_{ij_2}\|_\infty + \|b_{ij_1} b_{ij_2} - a_{ij_1} a_{ij_2}\|_{\dot{W}^{1,2d+2}}).$$

Clearly,

$$\|b_{ij_1} b_{ij_2} - a_{ij_1} a_{ij_2}\|_\infty$$
$$\leq \|b_{ij_1}\|_\infty \|b_{ij_2} - a_{ij_2}\|_\infty + \|a_{ij_2}\|_\infty \|b_{ij_1} - a_{ij_1}\|_\infty$$
$$\leq \|A - B\|_{L_\infty(\mathbb{H}^d, M_{2d}(\mathbb{R}))} (\|A\|_{L_\infty(\mathbb{H}^d, M_{2d}(\mathbb{R}))} + \|B\|_{L_\infty(\mathbb{H}^d, M_{2d}(\mathbb{R}))}).$$

Next,

$$\|X_k(b_{ij_1} b_{ij_2})\|_\infty \leq 2\|B\|_{L_\infty(\mathbb{H}^d, M_{2d}(\mathbb{R}))} \|B\|_{\dot{W}^{1,2d+2}(\mathbb{H}^d, M_{2d}(\mathbb{R}))}$$
$$\leq 2(\|A\|_{L_\infty(\mathbb{H}^d, M_{2d}(\mathbb{R}))}$$
$$+ \|B\|_{L_\infty(\mathbb{H}^d, M_{2d}(\mathbb{R}))}) \|B\|_{\dot{W}^{1,2d+2}(\mathbb{H}^d, M_{2d}(\mathbb{R}))}$$

and

$$\|X_k(a_{ij_1}a_{ij_2})\|_\infty \leq 2\|A\|_{L_\infty(\mathbb{H}^d, M_{2d}(\mathbb{R}))}\|A\|_{\dot{W}^{1,2d+2}(\mathbb{H}^d, M_{2d}(\mathbb{R}))}$$
$$\leq 2(\|A\|_{L_\infty(\mathbb{H}^d, M_{2d}(\mathbb{R}))}$$
$$+ \|B\|_{L_\infty(\mathbb{H}^d, M_{2d}(\mathbb{R}))})\|A\|_{\dot{W}^{1,2d+2}(\mathbb{H}^d, M_{2d}(\mathbb{R}))}.$$

Thus,

$$\|b_{ij_1}b_{ij_2} - a_{ij_1}a_{ij_2}\|_{\dot{W}^{1,2d+2}} \leq 2(\|A\|_{L_\infty(\mathbb{H}^d, M_{2d}(\mathbb{R}))} + \|B\|_{L_\infty(\mathbb{H}^d, M_{2d}(\mathbb{R}))}) \cdot$$
$$\cdot (\|A\|_{\dot{W}^{1,2d+2}(\mathbb{H}^d, M_{2d}(\mathbb{R}))} + \|B\|_{\dot{W}^{1,2d+2}(\mathbb{H}^d, M_{2d}(\mathbb{R}))}).$$

Thus,

$$\|T_{i,j_1,j_2}\|_\infty$$
$$\leq 2c'_d(\|A\|_{L_\infty(\mathbb{H}^d, M_{2d}(\mathbb{R}))} + \|B\|_{L_\infty(\mathbb{H}^d, M_{2d}(\mathbb{R}))}) \cdot$$
$$\cdot (\|A - B\|_{L_\infty(\mathbb{H}^d, M_{2d}(\mathbb{R}))} + \|A\|_{\dot{W}^{1,2d+2}(\mathbb{H}^d, M_{2d}(\mathbb{R}))} + \|B\|_{\dot{W}^{1,2d+2}(\mathbb{H}^d, M_{2d}(\mathbb{R}))}).$$

Substituting this inequality into (3.2.1), we complete the proof. $\square$

Proof of Theorem 3.2.4 We write

$$R_i^A - R_i^{A(p)} = (X_i^A - X_i^{A(p)})\Delta^{-\frac{1}{2}} \cdot \Delta^{\frac{1}{2}}(P^A)^{-\frac{1}{2}}$$
$$+ X_i^{A(p)}\Delta^{-\frac{1}{2}} \cdot \Delta^{\frac{1}{2}}(P^A)^{-\frac{1}{2}} \cdot ((P^{A(p)})^{\frac{1}{2}} - (P^A)^{\frac{1}{2}})(P^{A(p)})^{-\frac{1}{2}}.$$

Taking norms, we obtain

$$\|R_i^A - R_i^{A(p)}\|_\infty \leq \|(X_i^A - X_i^{A(p)})\Delta^{-\frac{1}{2}}\|_\infty \|\Delta^{\frac{1}{2}}(P^A)^{-\frac{1}{2}}\|_\infty$$
$$+ \|X_i^{A(p)}\Delta^{-\frac{1}{2}}\|_\infty \|\Delta^{\frac{1}{2}}(P^A)^{-\frac{1}{2}}\|_\infty \|((P^{A(p)})^{\frac{1}{2}} - (P^A)^{\frac{1}{2}})(P^{A(p)})^{-\frac{1}{2}}\|_\infty.$$

Using Theorem 3.1.7 and Lemma 3.2.5, we write

$$\|R_i^A - R_i^{A(p)}\|_\infty$$
$$\leq \|A - A(p)\|_{L_\infty(\mathbb{H}^d, M_{2d}(\mathbb{R}))}\|A^{-1}\|_{L_\infty(\mathbb{H}^d, M_{2d}(\mathbb{R}))} +$$
$$+ c_{\text{abs}}\|A\|_{L_\infty(\mathbb{H}^d, M_{2d}(\mathbb{R}))}\|A^{-1}\|_{L_\infty(\mathbb{H}^d, M_{2d}(\mathbb{R}))}$$
$$\times \|(P^A)^{-\frac{1}{4}}(P^A - P^{A(p)})(P^{A(u)})^{-\frac{3}{4}}\|_\infty.$$

(3.2.2)

3.2 Main Computational Theorem

Clearly,

$$(P^A)^{-\frac{1}{4}}(P^A - P^{A(p)})(P^{A(p)})^{-\frac{3}{4}}$$
$$= (P^A)^{-\frac{1}{4}}\Delta^{\frac{1}{4}} \cdot \Delta^{-\frac{1}{4}}(P^A - P^{A(p)})\Delta^{-\frac{3}{4}} \cdot \Delta^{\frac{3}{4}}(P^{A(p)})^{-\frac{3}{4}}.$$

Taking norms, we obtain

$$\|(P^A)^{-\frac{1}{4}}(P^A - P^{A(p)})(P^{A(p)})^{-\frac{3}{4}}\|_\infty$$
$$\leq \|(P^A)^{-\frac{1}{4}}\Delta^{\frac{1}{4}}\|_\infty \|\Delta^{-\frac{1}{4}}(P^A - P^{A(p)})\Delta^{-\frac{3}{4}}\|_\infty \|\Delta^{\frac{3}{4}}(P^{A(u)})^{-\frac{3}{4}}\|_\infty.$$

The first factor is controlled by Lemma 3.2.6. The second factor is controlled by Lemma 3.2.8. The third factor is controlled by Lemma 3.1.18. Thus,

$$\|(P^A)^{-\frac{1}{4}}(P^A - P^{A(p)})(P^{A(p)})^{-\frac{3}{4}}\|_\infty \tag{3.2.3}$$
$$\leq c_d \|A^{-1}\|^3_{L_\infty(\mathbb{H}^d, M_{2d}(\mathbb{R}))} \cdot$$
$$\cdot \|A\|^2_{L_\infty(\mathbb{H}^d, M_{2d}(\mathbb{R}))}(\|A - A(p)\|_{L_\infty(\mathbb{H}^d, M_{2d}(\mathbb{R}))} + \|A\|_{\dot{W}^{1,2d+2}(\mathbb{H}^d, M_{2d}(\mathbb{R}))}).$$

Combining (3.2.2) and (3.2.3), we complete the proof. $\square$

3.2.2 Compactness Theorems

Theorem 3.2.9 *Let $A, B : \mathbb{H}^d \to \mathrm{GL}(2d, \mathbb{R})$ be smooth functions, which are constant outside some ball. Let $\phi \in C_c^\infty(\mathbb{H}^d)$. Suppose $A = B$ in the neighborhood of $\mathrm{supp}(\phi)$. We have*

$$M_\phi R_i^A M_\phi - M_\phi R_i^B M_\phi \in \mathcal{K}(L_2(\mathbb{H}^d)).$$

Theorem 3.2.10 *Let $A : \mathbb{H}^d \to \mathrm{GL}(2d, \mathbb{R})$ be a smooth function, which is constant outside some ball. Let $\phi_1, \phi_2, \phi_3 \in C_c^\infty(\mathbb{H}^d)$. We have*

$$M_{\phi_1}[R_i^A, M_{\phi_2}]M_{\phi_3} \in \mathcal{K}(L_2(\mathbb{H}^d)).$$

The rest of this subsection is devoted to the proof of these theorems.

Lemma 3.2.11 *Let $A : \mathbb{H}^d \to \mathrm{GL}(2d, \mathbb{R})$ be a smooth function, which is constant outside some ball. Let $\phi \in C_c^\infty(\mathbb{H}^d)$. We have*

$$M_\phi R_i^A M_\phi - M_\phi X_i^A (1 + P^A)^{-\frac{1}{2}} M_\phi \in \mathcal{K}(L_2(\mathbb{H}^d)).$$

Proof We have

$$M_\phi R_i^A M_\phi - M_\phi X_i^A (1+P^A)^{-\frac{1}{2}} M_\phi$$

$$= M_\phi R_i^A \cdot \frac{1}{(1+P^A)^{\frac{1}{2}}((P^A)^{\frac{1}{2}}+(1+P^A)^{\frac{1}{2}})} \cdot M_\phi$$

$$= M_\phi R_i^A \cdot \frac{(P^A)^{\frac{1}{2}}}{(1+P^A)^{\frac{1}{2}}((P^A)^{\frac{1}{2}}+(1+P^A)^{\frac{1}{2}})} \cdot (P^A)^{-\frac{1}{2}} \Delta^{\frac{1}{2}} \cdot \Delta^{-\frac{1}{2}} M_\phi.$$

The first factor (on the right hand side) is bounded because R_i^A is bounded, the second factor is bounded by the functional calculus, the third factor is bounded by Lemma 3.2.5, the last factor belongs to $\mathcal{L}_{2d+2,\infty}$ by Cwikel estimate (see Theorem 1.1 in [48]). □

Lemma 3.2.12 *Let $A, B : \mathbb{H}^d \to \mathrm{GL}(2d, \mathbb{R})$ be smooth functions, which are constant outside some ball. Let $\phi \in C_c^\infty(\mathbb{H}^d)$. Suppose $A = B$ in a neighborhood of* supp(ϕ). *We have*

$$M_\phi X_i^A (1+P^A)^{-\frac{1}{2}} M_\phi - M_\phi X_i^B (1+P^B)^{-\frac{1}{2}} M_\phi \in \mathcal{K}(L_2(\mathbb{H}^d)).$$

Proof Clearly,

$$M_\phi X_i^A (1+P^A)^{-\frac{1}{2}} M_\phi - M_\phi X_i^B (1+P^B)^{-\frac{1}{2}} M_\phi$$

$$= M_\phi X_i^A \cdot \big((1+P^A)^{-\frac{1}{2}} - (1+P^B)^{-\frac{1}{2}}\big) \cdot M_\phi.$$

Recall the formula (valid for every positive self-adjoint x bounded from below by a strictly positive constant)

$$x^{-\frac{1}{2}} = \frac{2}{\pi} \int_0^\infty \frac{d\lambda}{x+\lambda^2}.$$

Thus,

$$(1+P^A)^{-\frac{1}{2}} - (1+P^B)^{-\frac{1}{2}}$$

$$= \frac{2}{\pi} \int_0^\infty \Big(\frac{1}{P^A+1+\lambda^2} - \frac{1}{P^B+1+\lambda^2}\Big) d\lambda$$

$$= \frac{2}{\pi} \int_0^\infty \frac{1}{P^A+1+\lambda^2} (P^B - P^A) \frac{1}{P^B+1+\lambda^2} d\lambda$$

3.2 Main Computational Theorem

and

$$M_\phi X_i^A (1+P^A)^{-\frac{1}{2}} M_\phi - M_\phi X_i^B (1+P^B)^{-\frac{1}{2}} M_\phi$$
$$= \frac{2}{\pi} \int_0^\infty M_\phi X_i^A \Big(\frac{1}{P^A+1+\lambda^2}(P^B - P^A)\frac{1}{P^B+1+\lambda^2}\Big) M_\phi \, d\lambda.$$

We now write $\phi = \phi_1 \phi_2$ with $\phi_1, \phi_2 \in C_c^\infty(\mathbb{H}^d)$ and $A = B$ in the neighborhood of $\text{supp}(\phi_1)$. We now write

$$\frac{1}{P^A+1+\lambda^2}(P^B - P^A)\frac{1}{P^B+1+\lambda^2} M_\phi$$
$$= \frac{1}{P^A+1+\lambda^2}(P^B - P^A) M_{\phi_1} \frac{1}{P^B+1+\lambda^2} M_{\phi_2}$$
$$+ \frac{1}{P^A+1+\lambda^2}(P^B - P^A)[\frac{1}{P^B+1+\lambda^2}, M_{\phi_1}] M_{\phi_2}.$$

Since $A = B$ in the neighborhood of $\text{supp}(\phi_1)$, it follows that $(P^A - P^B) M_{\phi_1} = 0$. Thus,

$$\frac{1}{P^A+1+\lambda^2}(P^B - P^A)\frac{1}{P^B+1+\lambda^2} M_\phi$$
$$= -\frac{1}{P^A+1+\lambda^2}(P^B - P^A)\frac{1}{P^B+1+\lambda^2}[P^B, M_{\phi_1}]\frac{1}{P^B+1+\lambda^2} M_{\phi_2}$$
$$= \frac{1}{P^B+1+\lambda^2}[P^B, M_{\phi_1}]\frac{1}{P^B+1+\lambda^2} M_{\phi_2}$$
$$- \frac{1}{P^A+1+\lambda^2}[P^B, M_{\phi_1}]\frac{1}{P^B+1+\lambda^2} M_{\phi_2}.$$

Consequently,

$$M_\phi X_i^A (1+P^A)^{-\frac{1}{2}} M_\phi - M_\phi X_i^B (1+P^B)^{-\frac{1}{2}} M_\phi$$
$$= M_\phi R_i^B \cdot \frac{2}{\pi} \int_0^\infty \frac{P^B}{P^B+1+\lambda^2} \cdot (P^B)^{-\frac{1}{2}}[P^B, M_{\phi_1}] \cdot \frac{1}{P^B+1+\lambda^2} M_{\phi_2} \, d\lambda$$
$$- M_\phi R_i^A \cdot \frac{2}{\pi} \int_0^\infty \frac{P^A}{P^A+1+\lambda^2} \cdot (P^A)^{-\frac{1}{2}}[P^B, M_{\phi_1}] \cdot \frac{1}{P^B+1+\lambda^2} M_{\phi_2} \, d\lambda.$$

Clearly, the integrands are compact and absolutely integrable in $\mathcal{L}_\infty$. Hence, the integrals are compact. □

Proof of Theorem 3.2.9 The assertion follows by combining Lemma 3.2.11 and Lemma 3.2.12. □

Lemma 3.2.13 *Let* $A : \mathbb{H}^d \to GL(2d, \mathbb{R})$ *be a smooth function, which is constant outside some ball. Let* $\phi \in C_c^\infty(\mathbb{H}^d)$. *We have*

$$[(1 + P^A)^{\frac{1}{2}}, M_\phi](1 + P^A)^{-\frac{1}{2}} \in \mathcal{K}(L_2(\mathbb{H}^d)).$$

Proof By Lemma 3.1.8, we have

$$x^{\frac{1}{2}} = \frac{2}{\pi} \int_0^\infty \frac{x}{x + \lambda^2} d\lambda.$$

Thus,

$$[(1 + P^A)^{\frac{1}{2}}, M_\phi](1 + P^A)^{-\frac{1}{2}} = \frac{2}{\pi} \int_0^\infty [\frac{1 + P^A}{1 + P^A + \lambda^2}, M_\phi](1 + P^A)^{-\frac{1}{2}} d\lambda.$$

Clearly,

$$[\frac{1 + P^A}{1 + P^A + \lambda^2}, M_\phi] = -\lambda^2 [\frac{1}{1 + P^A + \lambda^2}, M_\phi]$$

$$= \frac{\lambda^2}{1 + P^A + \lambda^2}[P^A, M_\phi]\frac{1}{1 + P^A + \lambda^2}.$$

Hence,

$$[\frac{1 + P^A}{1 + P^A + \lambda^2}, M_\phi](1 + P^A)^{-\frac{1}{2}}$$

$$= \frac{\lambda^2}{1 + P^A + \lambda^2} \cdot [P^A, M_\phi](1 + P^A)^{-\frac{1}{2}} \cdot \frac{1}{1 + P^A + \lambda^2}$$

$$= \frac{\lambda^2}{1 + P^A + \lambda^2} \cdot [P^A, M_\phi](1 + P^A)^{-\frac{3}{4}} \cdot \frac{(1 + P^A)^{\frac{1}{4}}}{1 + P^A + \lambda^2}.$$

In the right hand side, the first and third factors are bounded, while the second factor is compact. Hence, the integrand is compact. By Hölder inequality, we have

$$\|[\frac{1 + P^A}{1 + P^A + \lambda^2}, M_\phi](1 + P^A)^{-\frac{1}{2}}\|_\infty$$

$$\leq \|\frac{\lambda^2}{1 + P^A + \lambda^2}\|_\infty \|[P^A, M_\phi](1 + P^A)^{-\frac{1}{2}}\|_\infty \|\frac{1}{1 + P^A + \lambda^2}\|_\infty$$

$$\leq \frac{1}{1 + \lambda^2} \|[P^A, M_\phi](1 + P^A)^{-\frac{1}{2}}\|_\infty.$$

3.2 Main Computational Theorem

Hence, the integrand is also absolutely integrable in $\mathcal{L}_\infty$. Thus, the integral is compact and the assertion follows. □

Proof of Theorem 3.2.10 We have

$$[X_i^A(1+P^A)^{-\frac{1}{2}}, M_{\phi_2}] = [X_i^A, M_{\phi_2}](1+P^A)^{-\frac{1}{2}} - $$
$$- X_i^A(1+P^A)^{-\frac{1}{2}} \cdot [(1+P^A)^{\frac{1}{2}}, M_{\phi_2}](1+P^A)^{-\frac{1}{2}}.$$

The first summand is obviously compact. The second summand is compact by Lemma 3.2.13. Hence,

$$[X_i^A(1+P^A)^{-\frac{1}{2}}, M_{\phi_2}] \in \mathcal{K}(L_2(\mathbb{H}^d))$$

and

$$M_{\phi_1}[X_i^A(1+P^A)^{-\frac{1}{2}}, M_{\phi_2}]M_{\phi_3} \in \mathcal{K}(L_2(\mathbb{H}^d)).$$

However, Lemma 3.2.11 yields

$$M_{\phi_1}[R_i^A - X_i^A(1+P^A)^{-\frac{1}{2}}, M_{\phi_2}]M_{\phi_3} \in \mathcal{K}(L_2(\mathbb{H}^d)).$$

Combining these two equations, we complete the proof. □

3.2.3 Approximation Theorem

Let $\mathbb{H}^d(\mathbb{Z})$ be the natural lattice in $\mathbb{H}^d$. Define the group of shift operators by setting

$$(T_s f)(p) = f(ps^{-1}), \quad p, s \in \mathbb{H}^d.$$

Here, ps^{-1} is understood in the sense of $\mathbb{H}^d$.

Definition 3.2.14 Fix $\psi \in C_c(\mathbb{H}^d)$ be such that

$$\sum_{n \in \mathbb{H}^d(\mathbb{Z})} (T_n \psi)^2 = 1.$$

The sequence $\{T_n \psi\}_{n \in \mathbb{H}^d(\mathbb{Z})}$ is called a $\mathbb{H}^d(\mathbb{Z})$-invariant partition of unity.

Notation 3.2.15 For a function $\psi \in C_c(\mathbb{H}^d)$, we use the notation

$$\psi_{n,\epsilon} = (T_n \psi) \circ \delta_{\epsilon^{-1}}, \quad n \in \mathbb{H}^d(\mathbb{Z}), \quad \epsilon > 0.$$

Theorem 3.2.16 *Let $A : \mathbb{H}^d \to GL(2d, \mathbb{R})$ be a smooth function, which is constant outside some ball. Let $\psi \in C_c^\infty(\mathbb{H}^d)$ be as in Definition 3.2.14 and let $\psi(0) \neq 0$. For every $\epsilon > 0$, there exists a sequence $\{A_{n,\epsilon}\}_{n \in \mathbb{H}^d(\mathbb{Z})}$ of smooth functions $A_{n,\epsilon} : \mathbb{H}^d \to GL(2d, \mathbb{R})$ such that*

(i) $A_{n,\epsilon} = A$ on $\mathrm{supp}(\psi_{n,\epsilon})$.
(ii) $A_{n,\epsilon}$ *is a constant matrix outside some (dependent on n and ϵ) ball.*
(iii) *we have*

$$\lim_{\epsilon \downarrow 0} \sup_{n \in \mathbb{H}^d(\mathbb{Z})} \|A_{n,\epsilon} - A(\delta_\epsilon(n))\|_{L_\infty(\mathbb{H}^d, M_{2d}(\mathbb{R}))} = 0.$$

(iv) *we have*

$$\lim_{\epsilon \downarrow 0} \sup_{n \in \mathbb{H}^d(\mathbb{Z})} \|A_{n,\epsilon}\|_{\dot{W}^{1,2d+2}(\mathbb{H}^d, M_{2d}(\mathbb{R}))} = 0.$$

(v) *we have*

$$\sup_{\epsilon \in (0,1)} \sup_{n \in \mathbb{H}^d(\mathbb{Z})} \|A_{n,\epsilon}\|_{L_\infty(\mathbb{H}^d, M_{2d}(\mathbb{R}))} < \infty,$$

$$\sup_{\epsilon \in (0,1)} \sup_{n \in \mathbb{H}^d(\mathbb{Z})} \|A_{n,\epsilon}^{-1}\|_{L_\infty(\mathbb{H}^d, M_{2d}(\mathbb{R}))} < \infty.$$

Construction 3.2.17 Fix $r > 0$ and $\theta \in C_c^\infty(\mathbb{H}^d)$ compactly supported in $B(0, r)$ and set

$$\mathfrak{a}_{n,\epsilon}(p) = \log(A(p)A(\delta_\epsilon(n))^{-1}), \quad p \in B(\delta_\epsilon(n), r\epsilon),$$

$$A_{n,\epsilon}(p) = \exp(\theta_{n,\epsilon}(p)\mathfrak{a}_{n,\epsilon}(p)) \cdot A(\delta_\epsilon(n)), \quad p \in \mathbb{H}^d.$$

Here and below in this subsection, we denote by $B(p, r)$ the ball of radius r centered at p in $\mathbb{H}^d$ with respect to the Carnot-Caratheodory distance.

Our aim is to prove Theorem 3.2.16 exactly for this sequence $\{A_{n,\epsilon}\}_{n \in \mathbb{H}^d(\mathbb{Z})}$. The following assertion is standard (see e.g. Theorem 1.4 in [35]).

Lemma 3.2.18 *Let $A : \mathbb{H}^d \to GL(2d, \mathbb{R})$ be a smooth function. We have*

$$\|A(p_1) - A(p_2)\|_\infty \leq c_d \|A\|_{\dot{W}^{1,\infty}(\mathbb{H}^d, M_{2d}(\mathbb{R}))} \cdot \mathrm{dist}(p_1, p_2).$$

Here dist *is the Carnot-Caratheodory distance.*

3.2 Main Computational Theorem

Lemma 3.2.19 *Let $B : \Omega \to M_{2d}(\mathbb{R})$ be a smooth function.*

(i) We have
$$\|X_i(e^B)\|_\infty \leq e^{\|B\|_\infty} \|X_i B\|_\infty$$

(ii) If $\|B - 1\|_\infty \leq \frac{1}{2}$ everywhere on Ω, then also
$$\|X_i(\log(B))\|_\infty \leq 2\|X_i B\|_\infty$$

everywhere on Ω.

Proof We only prove (ii). The proof of (i) follows *mutatis mutandi*.

We have
$$\log(B) = \sum_{k \geq 0} \frac{(-1)^k}{k+1}(B-1)^{k+1}, \quad \|B-1\|_\infty < 1.$$

Thus,
$$X_i(\log(B)) = \sum_{k \geq 0} \frac{(-1)^k}{k+1} X_i((B-1)^{k+1}).$$

By the Leibniz rule, we have
$$X_i((B-1)^{k+1}) = \sum_{l=0}^{k} (B-1)^l \cdot X_i B \cdot (B-1)^{k-l}.$$

Thus,
$$\|X_i((B-1)^{k+1})\|_\infty \leq \sum_{l=0}^{k} \|(B-1)^l\|_\infty \|X_i B\|_\infty \|(B-1)^{k-l}\|_\infty \leq \frac{k+1}{2^k} \|X_i B\|_\infty.$$

By the triangle inequality, we have
$$\|X_i(\log(B))\|_\infty \leq \sum_{k \geq 0} \frac{1}{k+1} \cdot \frac{k+1}{2^k} \|X_i B\|_\infty = 2\|X_i B\|_\infty.$$

$\square$

Lemma 3.2.20 *There exists $\epsilon_0 > 0$ such that, for every $\epsilon \in (0, \epsilon_0)$ and for every $n \in \mathbb{H}^d(\mathbb{Z})$, $\mathfrak{a}_{n,\epsilon} : B(\delta_\epsilon(n), r\epsilon) \to M_{2d}(\mathbb{R})$ is a well-defined smooth mapping. We have*
$$\|\mathfrak{a}_{n,\epsilon}(p)\|_\infty \leq c_A r \epsilon \leq 1, \quad \|(X_i \mathfrak{a}_{n,\epsilon})(p)\|_\infty \leq c_A, \quad p \in B(\delta_\epsilon(n), r\epsilon). \tag{3.2.4}$$

Proof If $p \in B(\delta_\epsilon(n), r\epsilon)$, then

$$\|A(p)A(\delta_\epsilon(n))^{-1} - 1\|_\infty \leq \|A(p) - A(\delta_\epsilon(n))\|_\infty \|A(\delta_\epsilon(n))^{-1}\|_\infty$$

$$\leq \sup_{\substack{p_1, p_2 \in \mathbb{H}^d \\ \mathrm{dist}(p_1, p_2) \leq r\epsilon}} \|A(p_1) - A(p_2)\|_\infty \cdot \sup_{p_3 \in \mathbb{H}^d} \|A^{-1}(p_3)\|_\infty$$

$$\stackrel{L.3.2.18}{\leq} \frac{1}{2} c_A r\epsilon.$$

Set $\epsilon_0 = (c_A r)^{-1}$. It clearly follows from the first display that

$$\|A(p)A(\delta_\epsilon(n))^{-1} - 1\|_\infty \leq \frac{1}{2}, \quad p \in B(\delta_\epsilon(n), r\epsilon), \quad \epsilon \in (0, \epsilon_0).$$

If $X : \Omega \to M_{2d}(\mathbb{R})$ is a smooth mapping such that $\|X(p) - 1\|_\infty < 1$ for every $p \in \Omega$, then $\log(X) : \Omega \to M_{2d}(\mathbb{R})$ is also a smooth mapping. Hence, $\mathfrak{a}_{n,\epsilon} : B(\delta_\epsilon(n), r\epsilon) \to M_{2d}(\mathbb{R})$ is a well defined smooth mapping for every $\epsilon \in (0, \epsilon_0)$ and for every $n \in \mathbb{H}^d(\mathbb{Z})$.

The first inequality in (3.2.4) follows now from the elementary inequality

$$\|\log(X)\|_\infty \leq 2\|X - 1\|_\infty, \quad \|X - 1\|_\infty \leq \frac{1}{2}.$$

The second inequality in (3.2.4) follows now from Lemma 3.2.19 (ii). □

Proof of Theorem 3.2.16 Let $\theta \in C_c^\infty(\mathbb{H}^d)$ be such that $0 \leq \theta \leq 1$ and $\theta\psi = \psi$. Fix $r > 0$ such that θ is compactly supported in $B(0, r)$ (the ball is taken with respect to the Carnot-Caratheodory distance).

By Lemma 3.2.20, $\mathfrak{a}_{n,\epsilon} : B(\delta_\epsilon(n), r\epsilon) \to M_{2d}(\mathbb{R})$ is a well defined smooth mapping for every $\epsilon \in (0, \epsilon_0)$ and for every $n \in \mathbb{H}^d(\mathbb{Z})$. Hence, so is $A_{n,\epsilon} : B(\delta_\epsilon(n), r\epsilon) \to M_{2d}(\mathbb{R})$. Since θ vanishes near the boundary of $B(\delta_\epsilon(n), r\epsilon)$, it follows that $A_{n,\epsilon} = A(\delta_\epsilon(n))$ near the boundary of $B(\delta_\epsilon(n), r\epsilon)$. Since $A_{n,\epsilon} = A(\delta_\epsilon(n))$ outside $B(\delta_\epsilon(n), r\epsilon)$, the global smoothness of $A_{n,\epsilon}$ on $\mathbb{H}^d$ follows.

If $p \in \mathrm{supp}(\psi_{n,\epsilon})$, then $\theta_{n,\epsilon}(p) = 1$. It follows directly from Construction 3.2.17 that

$$A_{n,\epsilon}(p) = \exp(\mathfrak{a}_{n,\epsilon}(p)) \cdot A(\delta_\epsilon(n)) = A(p)A(\delta_\epsilon(n))^{-1} \cdot A(\delta_\epsilon(n)) = A(p).$$

This yields the assertion (i).

Since $A_{n,\epsilon} = A(\delta_\epsilon(n))$ outside $B(\delta_\epsilon(n), r\epsilon)$, the assertion (ii) follows.

Let us verify the assertion (iii). Recall that $0 \leq \theta \leq 1$. It follows from (3.2.4) and an elementary inequality

$$\|e^X - 1\|_\infty \leq 2\|X\|_\infty, \quad \|X\|_\infty \leq 1,$$

3.2 Main Computational Theorem

that
$$\|\exp(\theta_{n,\epsilon}(p)\mathfrak{a}_{n,\epsilon}(p)) - 1\|_\infty \leq 4c_A r\epsilon, \quad p \in B(\delta_\epsilon(n), r\epsilon).$$

If $p \notin B(\delta_\epsilon(n), r\epsilon)$, then the left hand side of the latter inequality is 0. Hence, the latter inequality holds for every $p \in \mathbb{H}^d$. We now have

$$\|A_{n,\epsilon}(p) - A(\delta_\epsilon(n))\|_\infty \leq \|\exp(\theta_{n,\epsilon}(p)\mathfrak{a}_{n,\epsilon}(p)) - 1\|_\infty \|A(\delta_\epsilon(n))\|_\infty$$
$$\leq 4c_A r\epsilon \cdot \sup_{p_1 \in \mathbb{H}^d} \|A(p_1)\|_\infty, \quad p \in \mathbb{H}^d.$$

This yields the assertion (iii).

Let us verify the assertion (iv). By the Leibniz rule, we have

$$X_i(\theta_{n,\epsilon} \mathfrak{a}_{n,\epsilon}) = X_i \theta_{n,\epsilon} \cdot \mathfrak{a}_{n,\epsilon} + \theta_{n,\epsilon} \cdot X_i \mathfrak{a}_{n,\epsilon}.$$

Thus,

$$\|X_i(\theta_{n,\epsilon} \mathfrak{a}_{n,\epsilon})\|_{L_{2d+2}(\mathbb{H}^d, M_{2d}(\mathbb{R}))} = \|X_i(\theta_{n,\epsilon} \mathfrak{a}_{n,\epsilon})\|_{L_{2d+2}(B(\delta_\epsilon(n), r\epsilon), M_{2d}(\mathbb{R}))}$$
$$\leq \|X_i \theta_{n,\epsilon}\|_{L_{2d+2}(B(\delta_\epsilon(n), r\epsilon))}$$
$$\cdot \|\mathfrak{a}_{n,\epsilon}\|_{L_\infty(B(\delta_\epsilon(n), r\epsilon), M_{2d}(\mathbb{R}))}$$
$$+ \|\theta_{n,\epsilon}\|_{L_{2d+2}(B(\delta_\epsilon(n), r\epsilon))}$$
$$\cdot \|X_i \mathfrak{a}_{n,\epsilon}\|_{L_\infty(B(\delta_\epsilon(n), r\epsilon), M_{2d}(\mathbb{R}))}$$
$$\leq \|X_i \theta\|_{L_{2d+2}(\mathbb{H}^d)} \cdot c_A r\epsilon + \epsilon \|\theta\|_{L_{2d+2}(\mathbb{H}^d)} \cdot c_A$$
$$= c_{A,\theta} \epsilon.$$

Since $0 \leq \theta \leq 1$, it follows from Lemma 3.2.19 that

$$\|X_i(\exp(\theta_{n,\epsilon} \mathfrak{a}_{n,\epsilon}))\|_{L_{2d+2}(\mathbb{H}^d, M_{2d}(\mathbb{R}))} \leq e \cdot c_{A,\theta} \epsilon.$$

Thus,

$$\|X_i(A_{n,\epsilon})\|_{L_{2d+2}(\mathbb{H}^d, M_{2d}(\mathbb{R}))} \leq \|X_i(\exp(\theta_{n,\epsilon} \mathfrak{a}_{n,\epsilon}))\|_{L_{2d+2}(\mathbb{H}^d, M_{2d}(\mathbb{R}))}$$
$$\cdot \|A(\delta_\epsilon(n))\|_\infty$$
$$\leq \|X_i(\exp(\theta_{n,\epsilon} \mathfrak{a}_{n,\epsilon}))\|_{L_{2d+2}(\mathbb{H}^d, M_{2d}(\mathbb{R}))}$$
$$\cdot \sup_{p_2 \in \mathbb{H}^d} \|A(p_2)\|_\infty$$
$$\leq e \cdot c_{A,\theta} \epsilon \cdot c'_A.$$

This yields the assertion (iv).

We have

$$\|A_{n,\epsilon}\|_{L_\infty(\mathbb{H}^d, M_{2d}(\mathbb{R}))} \leq \|\exp(\theta_{n,\epsilon} \mathfrak{a}_{n,\epsilon})\|_{L_\infty(\mathbb{H}^d, M_{2d}(\mathbb{R}))} \|A\|_{L_\infty(\mathbb{H}^d, M_{2d}(\mathbb{R}))}$$
$$\leq \exp(\|\theta_{n,\epsilon} \mathfrak{a}_{n,\epsilon}\|_{L_\infty(\mathbb{H}^d, M_{2d}(\mathbb{R}))}) \|A\|_{L_\infty(\mathbb{H}^d, M_{2d}(\mathbb{R}))}.$$

Similarly for $\|A_{n,\epsilon}^{-1}\|_{L_\infty(\mathbb{H}^d, M_{2d}(\mathbb{R}))}$. This yields the assertion (v). □

3.2.4 Proof of Theorem 3.2.2

Fix, throughout this section, $\phi \in C_c(\mathbb{H}^d)$. Fix a particular ψ as in Definition 3.2.14. Dependence on ϕ and ψ in various notations is suppressed.

Definition 3.2.21 Let $\epsilon > 0$. For $1 \leq i \leq 2d$, set

$$X_{i,\epsilon} = \sum_{n \in \mathbb{H}^d(\mathbb{Z})} M_{\psi_{n,\epsilon}} R_i^{A_{n,\epsilon}} M_{\psi_{n,\epsilon}}, \quad Y_{i,\epsilon} = \sum_{n \in \mathbb{H}^d(\mathbb{Z})} M_{\psi_{n,\epsilon}} R_i^{A(\delta_\epsilon(n))} M_{\psi_{n,\epsilon}}.$$

Lemma 3.2.22 Let $\phi \in C_c(\mathbb{H}^d)$. Let $\epsilon > 0$. For every $1 \leq i \leq 2d$, we have

$$M_\phi R_i^A M_\phi - M_\phi X_{i,\epsilon} M_\phi \in \mathcal{K}(L_2(\mathbb{H}^d)).$$

Proof Let $\mathscr{A}_\epsilon \subset \mathbb{H}^d(\mathbb{Z})$ consist of all $n \in \mathbb{H}^d(\mathbb{Z})$ such that

$$\psi_{n,\epsilon} \cdot \phi \neq 0.$$

Since ϕ and ψ are compactly supported, it follows that $\mathscr{A}_\epsilon$ is a finite set.

We write

$$M_\phi R_i^A M_\phi = \sum_{n \in \mathscr{A}_\epsilon} M_{\psi_{n,\epsilon}^2} M_\phi R_i^A M_\phi$$

$$= \sum_{n \in \mathscr{A}_\epsilon} M_{\psi_{n,\epsilon} \cdot \phi} R_i^A M_{\psi_{n,\epsilon} \cdot \phi} + \sum_{n \in \mathscr{A}_\epsilon} M_{\psi_{n,\epsilon} \cdot \phi} [M_{\psi_{n,\epsilon}}, R_i^A] M_\phi$$

$$= X_{i,\epsilon} + \sum_{n \in \mathscr{A}_\epsilon} M_{\psi_{n,\epsilon} \cdot \phi} (R_i^A - R_i^{A_{n,\epsilon}}) M_{\psi_{n,\epsilon} \cdot \phi}$$

$$+ \sum_{n \in \mathscr{A}_\epsilon} M_{\psi_{n,\epsilon} \cdot \phi} [M_{\psi_{n,\epsilon}}, R_i^A] M_\phi.$$

The summands in the first sum (on the right hand side) are compact by Theorem 3.2.9. Since $\mathscr{A}_\epsilon$ is finite, it follows that the first sum is compact. The summands

3.2 Main Computational Theorem

in the second sum are compact by Theorem 3.2.10. Since $\mathscr{A}_\epsilon$ is a finite set, it follows that the second sum is compact. This completes the proof. □

Lemma 3.2.23 *Let* $K \subset \mathbb{H}^d$ *be a bounded set. Let* $\{T_n\}_{n \in \mathbb{H}^d(\mathbb{Z})}$ *be such that* $T_n T_m = T_n^* T_m = 0$ *unless* $mn^{-1} \in K$. *We have*

$$\| \sum_{n \in \mathbb{H}^d(\mathbb{Z})} T_n \|_\infty \leq c_K \sup_{n \in \mathbb{H}^d(\mathbb{Z})} \|T_n\|_\infty.$$

Proof Choose $l \in \mathbb{N}$ such that $mn^{-1} \notin K$ for $m, n \in \delta_l(\mathbb{H}^d(\mathbb{Z}))$, $m \neq n$. Let **n** be an element of the quotient space $\delta_l(\mathbb{H}^d(\mathbb{Z}))/\mathbb{H}^d(\mathbb{Z})$. We have

$$T_n T_m = T_n^* T_m = 0, \quad m, n \in \mathbf{n}, \quad m \neq n.$$

We write

$$\sum_{n \in \mathbb{H}^d(\mathbb{Z})} T_n = \sum_{\mathbf{n} \in \delta_l(\mathbb{H}^d(\mathbb{Z}))/\mathbb{H}^d(\mathbb{Z})} \sum_{n \in \mathbf{n}} T_n.$$

By triangle inequality,

$$\| \sum_{n \in \mathbb{H}^d(\mathbb{Z})} T_n \|_\infty \leq \sum_{\mathbf{n} \in \delta_l(\mathbb{H}^d(\mathbb{Z}))/\mathbb{H}^d(\mathbb{Z})} \| \sum_{n \in \mathbf{n}} T_n \|_\infty.$$

For every $\mathbf{n} \in \delta_l(\mathbb{H}^d(\mathbb{Z}))/\mathbb{H}^d(\mathbb{Z})$, the operators $\{T_n\}_{n \in \mathbf{n}}$ are pairwise orthogonal. Thus,

$$\| \sum_{n \in \mathbf{n}} T_n \|_\infty = \sup_{n \in \mathbf{n}} \|T_n\|_\infty.$$

We, therefore, have

$$\| \sum_{n \in \mathbb{Z}^{2d+1}} T_n \|_\infty \leq \sum_{\mathbf{n} \in \delta_l(\mathbb{H}^d(\mathbb{Z}))/\mathbb{H}^d(\mathbb{Z})} \sup_{n \in \mathbf{n}} \|T_n\|_\infty.$$

Since the quotient space is finite, the assertion follows immediately. □

Lemma 3.2.24 *We have* $X_{i,\epsilon} - Y_{i,\epsilon} \to 0$ *in the uniform norm as* $\epsilon \downarrow 0$.

Proof By Definition 3.2.21 and Theorem 3.2.16 (i), we have

$$X_{i,\epsilon} - Y_{i,\epsilon} = \sum_{n \in \mathbb{H}^d(\mathbb{Z})} M_{\psi_{n,\epsilon}} (R_i^{A_{n,\epsilon}} - R_i^{A_{n,\epsilon}(\delta_\epsilon(n))}) M_{\psi_{n,\epsilon}}.$$

By Lemma 3.2.23, we have

$$\|X_{i,\epsilon} - Y_{i,\epsilon}\|_\infty \leq c_\psi \sup_{n\in\mathbb{H}^d(\mathbb{Z})} \|M_{\psi_{n,\epsilon}}(R_i^{A_{n,\epsilon}} - R_i^{A_{n,\epsilon}(\delta_\epsilon(n))})M_{\psi_{n,\epsilon}}\|_\infty$$

$$\leq c_\psi \sup_{n\in\mathbb{H}^d(\mathbb{Z})} \|R_i^{A_{n,\epsilon}} - R_i^{A_{n,\epsilon}(\delta_\epsilon(n))}\|_\infty.$$

Using Theorem 3.2.4, we write

$$\|X_{i,\epsilon} - Y_{i,\epsilon}\|_\infty$$
$$\leq c_d c_\psi \big(\sup_{n\in\mathbb{H}^d(\mathbb{Z})} \|A_{n,\epsilon} - A(\delta_\epsilon(n))\|_{L_\infty(\mathbb{H}^d, M_{2d}(\mathbb{R}))}$$
$$+ \sup_{n\in\mathbb{H}^d(\mathbb{Z})} \|A_{n,\epsilon}\|_{\dot{W}^{1,2d+2}(\mathbb{H}^d, M_{2d}(\mathbb{R}))} \big)$$
$$\cdot \sup_{n\in\mathbb{H}^d(\mathbb{Z})} \|A_{n,\epsilon}\|^3_{L_\infty(\mathbb{H}^d, M_{2d}(\mathbb{R}))} \cdot \sup_{n\in\mathbb{H}^d(\mathbb{Z})} \|A_{n,\epsilon}^{-1}\|^4_{L_\infty(\mathbb{H}^d, M_{2d}(\mathbb{R}))}.$$

The assertion follows now from Theorem 3.2.16. □

Recall that we identify the symbol algebra $(C_0+\mathbb{C})(\mathbb{H}^d) \otimes_{\min} C^*(\{R_k\}_{k=1}^{2d})$ with the algebra $(C_0+\mathbb{C})(\mathbb{H}^d, C^*(\{R_k\}_{k=1}^{2d}))$.

Lemma 3.2.25 *Let $\phi \in C_c(\mathbb{H}^d)$. For every $\epsilon > 0$, $M_\phi Y_{i,\epsilon} M_\phi \in \Pi$ and*

$$\lim_{\epsilon \downarrow 0} \big(\mathrm{sym}(M_\phi Y_{i,\epsilon} M_\phi)\big)(p) = \phi^2(p) \cdot R_i^{A(p)}, \quad p \in \mathbb{H}^d.$$

Proof Let a finite set $\mathscr{A}_\epsilon \subset \mathbb{H}^d(\mathbb{Z})$ consist of all $n \in \mathbb{H}^d(\mathbb{Z})$ such that

$$\psi_{n,\epsilon} \cdot \phi \neq 0.$$

It follows from Definition 3.2.21 that

$$M_\phi Y_{i,\epsilon} M_\phi = \sum_{n\in\mathscr{A}_\epsilon} M_{\phi\cdot\psi_{n,\epsilon}} R_i^{A(\delta_\epsilon(n))} M_{\psi_{n,\epsilon}\cdot\phi}.$$

Noting that $\mathscr{A}_\epsilon$ is a finite set, we obtain $M_\phi Y_{i,\epsilon} M_\phi \in \Pi$. We have

$$\mathrm{sym}(M_\phi Y_{i,\epsilon} M_\phi) = \sum_{n\in\mathscr{A}_\epsilon} \phi^2 \psi_{n,\epsilon}^2 \otimes R_i^{A(\delta_\epsilon(n))} = \sum_{n\in\mathbb{H}^d(\mathbb{Z})} \phi^2 \psi_{n,\epsilon}^2 \otimes R_i^{A(\delta_\epsilon(n))}.$$

3.2 Main Computational Theorem

For every $p \in \mathbb{H}^d$, we have

$$\mathrm{sym}(M_\phi Y_{i,\epsilon} M_\phi)(p) = \phi(p)^2 \sum_{n \in \mathbb{H}^d(\mathbb{Z})} \psi_{n,\epsilon}^2(p) R_i^{A(\delta_\epsilon(n))}$$

$$= \phi^2(p) R_i^{A(p)} + \phi(p)^2 \sum_{n \in \mathbb{H}^d(\mathbb{Z})} \psi_{n,\epsilon}^2(p)(R_i^{A(\delta_\epsilon(n))} - R_i^{A(p)}).$$

Thus,

$$\|\mathrm{sym}(M_\phi Y_{i,\epsilon} M_\phi)(p) - \phi^2(t) R_i^{A(p)}\|_\infty \leq \sum_{n \in \mathbb{H}^d(\mathbb{Z})} \psi_{n,\epsilon}^2(p) \|R_i^{A(\delta_\epsilon(n))} - R_i^{A(p)}\|_\infty$$

$$\leq \sup_{\substack{n \in \mathbb{H}^d(\mathbb{Z}) \\ \psi_{n,\epsilon}(p) \neq 0}} \|R_i^{A(\delta_\epsilon(n))} - R_i^{A(p)}\|_\infty.$$

Using Theorem 3.1.12, we write

$$\|\mathrm{sym}(M_\phi Y_{i,\epsilon} M_\phi)(p) - \phi^2(t) R_i^{A(p)}\|_\infty$$

$$\leq c_d \sup_{\substack{n \in \mathbb{H}^d(\mathbb{Z}) \\ \psi_{n,\epsilon}(p) \neq 0}} \|A(p) - A(\delta_\epsilon(n))\|_\infty \cdot \|A\|^3_{L_\infty(\mathbb{H}^d, M_{2d}(\mathbb{R}))} \cdot \|A^{-1}\|^4_{L_\infty(\mathbb{H}^d, M_{2d}(\mathbb{R}))}.$$

Since the right hand side clearly tends to 0 as $\epsilon \downarrow 0$, the assertion follows. □

Proof of Theorem 3.2.2 By Lemma 3.2.22 and Theorem 1.3.2, we have

$$M_\phi R_i^A M_\phi - M_\phi X_{i,\epsilon} M_\phi = K_{i,\epsilon} \in \mathcal{K}(L_2(\mathbb{H}^d)) \subset \Pi.$$

We now write

$$M_\phi R_i^A M_\phi = (K_{i,\epsilon} + M_\phi Y_{i,\epsilon} M_\phi) + M_\phi(X_{i,\epsilon} - Y_{i,\epsilon}) M_\phi.$$

By Lemma 3.2.24, the second summand tends to 0 in the uniform norm as $\epsilon \downarrow 0$. The first summand, clearly, belongs to Π by Theorem 1.3.2 and Lemma 3.2.25. It follows that $M_\phi R_i^A M_\phi$ belongs to the closure of Π in the uniform norm and, therefore, to Π.

Furthermore, we have

$$\mathrm{sym}(M_\phi R_i^A M_\phi) = \lim_{\epsilon \downarrow 0} \mathrm{sym}(K_{i,\epsilon} + M_\phi Y_{i,\epsilon} M_\phi) = \lim_{\epsilon \downarrow 0} \mathrm{sym}(M_\phi Y_{i,\epsilon} M_\phi).$$

Here, the convergence is understood in the norm and, therefore, pointwise. The assertion follows now from Lemma 3.2.25. □

3.3 Invariance of Principal Symbol Map

3.3.1 Conjugation of the Differential Calculus by the Heisenberg Diffeomorphisms

The following conjugation result is standard, however we prove it for convenience of the reader.

Theorem 3.3.1 *Let $\Omega, \Omega' \subset \mathbb{H}^d$ be open subsets and let $\Phi : \Omega \to \Omega'$ be a Heisenberg diffeomorphism. Let $V_\Phi \xi = \xi \circ \Phi$. We have (on Ω')*

$$V_\Phi^{-1} X_j V_\Phi = \sum_{l=1}^{2d} M_{X_j \Phi_l \circ \Phi^{-1}} X_l, \quad 1 \leq j \leq 2d.$$

Recall that $p \in \mathbb{H}^d$ has a coordinate representation $p = (x_1, \cdots, x_d, y_1, \cdots, y_d, t)$. We also need a coordinate representation $\Phi = (\Phi_1, \cdots, \Phi_{2d+1})$ of a Heisenberg diffeomorphism Φ.

Lemma 3.3.2 *Let $\Phi : \mathbb{H}^d \to \mathbb{H}^d$ be a Heisenberg diffeomorphism. We have*

$$\sum_{l=1}^{d} (y_l \circ \Phi) \cdot X_j \Phi_l - \sum_{l=d+1}^{2d} (x_{l-d} \circ \Phi) \cdot X_j \Phi_l + X_j \Phi_{2d+1} = 0, \quad 1 \leq j \leq 2d.$$

Proof Fix $p = (x_1, \cdots, x_d, y_1, \cdots, y_d, t)$. Set

$$q^j = \begin{cases} e_j - y_j e_{2d+1}, & 1 \leq j \leq d \\ e_j + x_{j-d} e_{2d+1}, & d+1 \leq j \leq 2d \end{cases}$$

Note that $q^j \in F(p)^\perp$.

By the definition of the Jacobi matrix, we have

$$\iota\{(D_j \Phi_l)(p)\}_{l=1}^{2d+1} = J_\Phi(p) e_j, \quad 1 \leq j \leq 2d+1.$$

For $1 \leq j \leq d$, we write

$$\{(X_j \Phi_l)(p)\}_{l=1}^{2d+1} = \iota\{(D_j \Phi_l)(p)\}_{l=1}^{2d+1} - \iota y_j \cdot \{(D_{2d+1} \Phi_l)(p)\}_{l=1}^{2d+1}$$
$$= J_\Phi(p) e_j - y_j \cdot J_\phi(p) e_{2d+1}$$
$$= J_\Phi(p) q^j.$$

3.3 Invariance of Principal Symbol Map

For $d+1 \leq j \leq 2d$, we have

$$\{(X_j\Phi_l)(p)\}_{l=1}^{2d+1} = \iota\{(D_j\Phi_l)(p)\}_{l=1}^{2d+1} + \iota x_j \cdot \{(D_{2d+1}\Phi_l)(p)\}_{l=1}^{2d+1}$$
$$= J_\Phi(p)e_j + x_j \cdot J_\Phi(p)e_{2d+1}$$
$$= J_\Phi(p)q^j.$$

Using the definition of F (see Example 1.3.3), we write

$$\left(\sum_{l=1}^{d}(y_l \circ \Phi) \cdot X_j\Phi_l - \sum_{l=d+1}^{2d}(x_{l-d} \circ \Phi) \cdot X_j\Phi_l + X_j\Phi_{2d+1}\right)(p)$$
$$= \langle F(\Phi(p)), \{(X_j\Phi_l)(p)\}_{l=1}^{2d+1}\rangle$$
$$= \langle F(\Phi(p)), J_\Phi(p)q^j\rangle.$$

Since $q^j \in F(p)^\perp$, the assertion follows directly from the definition of the Heisenberg diffeomorphism. □

Proof of Theorem 3.3.1 By the chain rule, we have

$$D_k V_\Phi \xi = D_k(\xi \circ \Phi) = \sum_{l=1}^{2d+1}((D_l\xi) \circ \Phi) \cdot \iota D_k \Phi_l.$$

Therefore,

$$X_j V_\Phi \xi = \sum_{l=1}^{2d+1}((\iota D_l\xi) \circ \Phi) \cdot X_j\Phi_l.$$

Recall that

$$\iota D_l = \begin{cases} X_l + M_{y_l}T, & 1 \leq l \leq d \\ X_l - M_{x_{l-d}}T, & d+1 \leq l \leq 2d \\ T, & l = 2d+1 \end{cases}$$

Thus,

$$X_j V_\Phi \xi = \sum_{l=1}^{2d}(X_l\xi \circ \Phi) \cdot X_j\Phi_l + (T\xi \circ \Phi) \cdot a_j,$$

where

$$a_j = \sum_{l=1}^{d}(y_l \circ \Phi) \cdot X_j\Phi_l - \sum_{l=d+1}^{2d}(x_{l-d} \circ \Phi) \cdot X_j\Phi_l + X_j\Phi_{2d+1} \stackrel{L.3.3.2}{=} 0.$$

It follows that
$$X_j V_\Phi \xi = \sum_{l=1}^{2d} (X_l \xi \circ \Phi) \cdot X_j \Phi_l.$$

Since ξ is arbitrary, it follows that
$$V_\Phi^{-1} X_j V_\Phi = \sum_{l=1}^{2d} M_{X_j \Phi_l \circ \Phi^{-1}} X_l.$$

□

The next lemma follows immediately from Theorem 3.3.1 and the Leibniz rule.

Lemma 3.3.3 *Let $\Phi : \mathbb{H}^d \to \mathbb{H}^d$ be a Heisenberg diffeomorphism. Recall the notation*
$$(U_\Phi \xi)(p) = |\det(J_\Phi)(p)|^{\frac{1}{2}} \xi(\Phi(p)), \quad \xi \in L_2(\mathbb{H}^d), \quad p \in \mathbb{H}^d.$$

We have
$$U_\Phi^{-1} X_j U_\Phi = X_j^{(HJ_\Phi)^* \circ \Phi^{-1}} + M_{a_j}, \quad 1 \le j \le 2d,$$

with some $a_j \in C^\infty(\mathbb{H}^d)$, $1 \le j \le 2d$.

3.3.2 Equivariance Result on the Heisenberg Group

The main result of this subsection is the following theorem. It explains our motivation to consider operators of the form R^A with A being a function on $\mathbb{H}^d$.

Theorem 3.3.4 *Let $\phi \in C_c(\mathbb{H}^d)$ and let $\Phi : \mathbb{H}^d \to \mathbb{H}^d$ be a diffeomorphism. Let $A : \mathbb{H}^d \to \mathrm{Aut}(\mathbb{H}^d)$ be a smooth mapping. Suppose that*

(i) *Φ is Heisenberg in a neighborhood of $\mathrm{supp}(\phi \circ \Phi)$;*
(ii) *Φ is affine outside of some ball;*
(iii) *A and A^{-1} are bounded;*
(iv) *$A = A^\Phi \circ \Phi^{-1}$ in a neighborhood of $\mathrm{supp}(\phi)$.*

We have
$$M_\phi \left(U_\Phi^{-1} R_i U_\Phi - R_i^A \right) M_\phi \in \mathcal{K}(L_2(\mathbb{H}^d)).$$

The rest of this subsection is devoted to the proof of this theorem.

3.3 Invariance of Principal Symbol Map

Lemma 3.3.5 *In the setting of Theorem 3.3.4, we have*

$$\left((1 - U_\Phi^{-1}\Delta U_\Phi)^{-\frac{1}{2}} - (1 + P^A)^{-\frac{1}{2}}\right)M_\phi$$

$$= \int_0^\infty U_\Phi^{-1}(1 + \lambda^2 + \Delta)^{-1} U_\Phi T_1 \frac{(P^A)^{\frac{1}{2}}}{1 + \lambda^2 + P^A} M_{\phi_2} d\lambda$$

$$+ \int_0^\infty (1 + \lambda^2 + P^A)^{-1} T_2 \frac{(P^A)^{\frac{1}{2}}}{1 + \lambda^2 + P^A} M_{\phi_2} d\lambda.$$

Here, T_1 and T_2 are some bounded operators.

Proof Recall the formula

$$x^{-\frac{1}{2}} = \frac{2}{\pi} \int_0^\infty \frac{d\lambda}{x + \lambda^2}, \quad x \geq 0.$$

We write

$$\left((1 - U_\Phi^{-1}\Delta U_\Phi)^{-\frac{1}{2}} - (1 + P^A)^{-\frac{1}{2}}\right)M_\phi$$

$$= \frac{2}{\pi} \int_0^\infty \left((1 + \lambda^2 - U_\Phi^{-1}\Delta U_\Phi)^{-1} - (1 + \lambda^2 + P^A)^{-1}\right) M_\phi d\lambda.$$

Write $\phi = \phi_1 \phi_2$, $\phi_1 \in C_c^\infty(\mathbb{H}^d)$, $\phi_2 \in C_c(\mathbb{H}^d)$, $\mathrm{supp}(\phi_1) = \mathrm{supp}(\phi)$. Clearly,

$$Q_1 = (P^A + U_\Phi^{-1}\Delta U_\Phi) M_{\phi_1}$$

is a differential operator of order 1 with smooth compactly supported coefficients. Also denote

$$Q_2 = [P^A, M_{\phi_1}].$$

This is also a differential operator with smooth compactly supported coefficients. Then we proceed as follows:

$$\left((1 + \lambda^2 - U_\Phi^{-1}\Delta U_\Phi)^{-1} - (1 + \lambda^2 + P^A)^{-1}\right) M_\phi$$

$$= (1 + \lambda^2 - U_\Phi^{-1}\Delta U_\Phi)^{-1}(P^A + U_\Phi^{-1}\Delta U_\Phi)(1 + \lambda^2 + P^A)^{-1} M_\phi$$

$$= (1 + \lambda^2 - U_\Phi^{-1}\Delta U_\Phi)^{-1}(P^A + U_\Phi^{-1}\Delta U_\Phi) M_{\phi_1}(1 + \lambda^2 + P^A)^{-1} M_{\phi_2} +$$

$$+ (1 + \lambda^2 - U_\Phi^{-1}\Delta U_\Phi)^{-1}(P^A + U_\Phi^{-1}\Delta U_\Phi)[(1 + \lambda^2 + P^A)^{-1}, M_{\phi_1}] M_{\phi_2}$$

$$=(1+\lambda^2 - U_\Phi^{-1}\Delta U_\Phi)^{-1} \cdot Q\Delta^{-\frac{1}{2}} \cdot \Delta^{\frac{1}{2}}(P^A)^{-\frac{1}{2}} \cdot \frac{(P^A)^{\frac{1}{2}}}{1+\lambda^2+P^A} M_{\phi_2} +$$
$$- (1+\lambda^2 - U_\Phi^{-1}\Delta U_\Phi)^{-1}(P^A + U_\Phi^{-1}\Delta U_\Phi)(1+\lambda^2+P^A)^{-1}[P^A, M_{\phi_1}] \times$$
$$\times (1+\lambda^2+P^A)^{-1} M_{\phi_2}$$

$$=(1+\lambda^2 - U_\Phi^{-1}\Delta U_\Phi)^{-1} \cdot Q_1\Delta^{-\frac{1}{2}} \cdot \Delta^{\frac{1}{2}}(P^A)^{-\frac{1}{2}} \cdot \frac{(P^A)^{\frac{1}{2}}}{1+\lambda^2+P^A} M_{\phi_2} +$$
$$- (1+\lambda^2 - U_\Phi^{-1}\Delta U_\Phi)^{-1}[P^A, M_{\phi_1}](1+\lambda^2+P^A)^{-1} M_{\phi_2} +$$
$$+ (1+\lambda^2+P^A)^{-1}[P^A, M_{\phi_1}](1+\lambda^2+P^A)^{-1} M_{\phi_2}$$

$$=(1+\lambda^2 - U_\Phi^{-1}\Delta U_\Phi)^{-1} \cdot Q_1\Delta^{-\frac{1}{2}} \cdot \Delta^{\frac{1}{2}}(P^A)^{-\frac{1}{2}} \cdot \frac{(P^A)^{\frac{1}{2}}}{1+\lambda^2+P^A} M_{\phi_2} +$$
$$- (1+\lambda^2 - U_\Phi^{-1}\Delta U_\Phi)^{-1} \cdot Q_2\Delta^{-\frac{1}{2}} \cdot \Delta^{\frac{1}{2}}(P^A)^{-\frac{1}{2}} \cdot \frac{(P^A)^{\frac{1}{2}}}{1+\lambda^2+P^A} M_{\phi_2} +$$
$$+ (1+\lambda^2+P^A)^{-1} \cdot Q_2\Delta^{-\frac{1}{2}} \cdot \Delta^{\frac{1}{2}}(P^A)^{-\frac{1}{2}} \cdot \frac{(P^A)^{\frac{1}{2}}}{1+\lambda^2+P^A} M_{\phi_2}.$$

Setting

$$T_1 = \frac{2}{\pi} Q_1 \Delta^{-\frac{1}{2}} \cdot \Delta^{\frac{1}{2}}(P^A)^{-\frac{1}{2}} - \frac{2}{\pi} Q_2 \Delta^{-\frac{1}{2}} \cdot \Delta^{\frac{1}{2}}(P^A)^{-\frac{1}{2}},$$

$$T_2 = \frac{2}{\pi} Q_2 \Delta^{-\frac{1}{2}} \cdot \Delta^{\frac{1}{2}}(P^A)^{-\frac{1}{2}},$$

we complete the proof. □

Lemma 3.3.6 *In the setting of Theorem 3.3.4, we have*

$$M_\phi U_\Phi^{-1} X_i U_\Phi \left((1 - U_\Phi^{-1}\Delta U_\Phi)^{-\frac{1}{2}} - (1+P^A)^{-\frac{1}{2}}\right) M_\phi \in \mathcal{K}(L_2(\mathbb{H}^d)).$$

Proof The operator $M_\phi U_\Phi^{-1} X_i U_\Phi$ is a first order differential operator with smooth compactly supported coefficients. Denote

$$T_3 = M_\phi U_\Phi^{-1} X_i U_\Phi \Delta^{-\frac{1}{2}} \cdot \Delta^{\frac{1}{2}}(P^A)^{-\frac{1}{2}}.$$

3.3 Invariance of Principal Symbol Map

Clearly, T_3 is a bounded operator. By Lemma 3.3.5, we have

$$M_\phi U_\Phi^{-1} X_i U_\Phi \left((1 - U_\Phi^{-1} \Delta U_\Phi)^{-\frac{1}{2}} - (1 + P^A)^{-\frac{1}{2}} \right) M_\phi$$

$$= \int_0^\infty U_\Phi^{-1} X_i (1 + \lambda^2 + \Delta)^{-1} U_\Phi T_1 \frac{(P^A)^{\frac{1}{2}}}{1 + \lambda^2 + P^A} M_{\phi_2} d\lambda +$$

$$+ T_3 \cdot \int_0^\infty \frac{(P^A)^{\frac{1}{2}}}{1 + \lambda^2 + P^A} T_2 \frac{(P^A)^{\frac{1}{2}}}{1 + \lambda^2 + P^A} M_{\phi_2} d\lambda.$$

Since

$$\frac{((P^A)^{\frac{1}{2}}}{1 + \lambda^2 + P^A} M_{\phi_2} = \frac{P^A}{1 + \lambda^2 + P^A} \cdot (P^A)^{-\frac{1}{2}} \Delta^{\frac{1}{2}} \cdot \Delta^{-\frac{1}{2}} M_{\phi_2}$$

is compact, it follows that both integrands are compact operators.

Note that

$$\left\| U_\Phi^{-1} X_i (1 + \lambda^2 + \Delta)^{-1} U_\Phi T_1 \frac{(P^A)^{\frac{1}{2}}}{1 + \lambda^2 + P^A} M_{\phi_2} \right\|_\infty$$

$$\leq \| X_i (1 + \lambda^2 + \Delta)^{-1} \|_\infty \| T_1 \|_\infty \left\| \frac{(P^A)^{\frac{1}{2}}}{1 + \lambda^2 + P^A} \right\|_\infty \| \phi_2 \|_\infty$$

$$\leq \sup_{s \geq 1} \frac{s}{s^2 + \lambda^2} \cdot \| T_1 \|_\infty \cdot \sup_{s \geq 1} \frac{s}{s^2 + \lambda^2} \cdot \| \phi_2 \|_\infty$$

$$\leq \min\{1, \lambda^{-2}\} \cdot \| T_1 \|_\infty \| \phi_2 \|_\infty.$$

Similarly,

$$\left\| \frac{(P^A)^{\frac{1}{2}}}{1 + \lambda^2 + P^A} T_2 \frac{(P^A)^{\frac{1}{2}}}{1 + \lambda^2 + P^A} M_{\phi_2} \right\|_\infty \leq \min\{1, \lambda^{-2}\} \cdot \| T_2 \|_\infty \| \phi_2 \|_\infty.$$

Hence, both integrands are absolutely integrable in $\mathcal{K}(L_2(\mathbb{H}^d))$. This completes the proof. □

Proof of Theorem 3.3.4 We write

$$M_\phi \left(U_\Phi^{-1} R_i U_\Phi - R_i^A \right) M_\phi$$

$$= M_\phi U_\Phi^{-1} (X_i \Delta^{-\frac{1}{2}} - X_i (1 + \Delta)^{-\frac{1}{2}}) U_\Phi M_\phi +$$

$$+ M_\phi U_\Phi^{-1} X_i U_\Phi \left((1 - U_\Phi^{-1} \Delta U_\Phi)^{-\frac{1}{2}} - (1 + P^A)^{-\frac{1}{2}} \right) M_\phi +$$

$$+ M_\phi (U_\Phi^{-1} X_i U_\Phi - X_i^A)(1 + P^A)^{-\frac{1}{2}} M_\phi + M_\phi X_i^A ((1 + P^A)^{-\frac{1}{2}} - (P^A)^{-\frac{1}{2}}) M_\phi.$$

The first summand on the right hand side is

$$M_\phi U_\Phi^{-1} R_i U_\Phi \cdot U_\Phi^{-1}\left(1 - \frac{\Delta^{\frac{1}{2}}}{(1+\Delta)^{\frac{1}{2}}}\right) M_{\phi \circ \Phi} U_\Phi.$$

Since

$$\left(1 - \frac{\Delta^{\frac{1}{2}}}{(1+\Delta)^{\frac{1}{2}}}\right) M_{\phi \circ \Phi}$$

is compact, it follows that so is the first summand.

The second summand is compact by Lemma 3.3.6.

By the assumption on A, we have that $M_\phi(U_\Phi^{-1} X_i U_\Phi - X_i^A) = M_\theta$ with some $\theta \in C_c^\infty(\mathbb{H}^d)$. Hence, the third summand is written as

$$M_\theta \Delta^{-\frac{1}{2}} \cdot \Delta^{\frac{1}{2}} (P^A)^{-\frac{1}{2}} \cdot (P^A)^{\frac{1}{2}} (1 + P^A)^{-\frac{1}{2}} \cdot M_\phi.$$

It follows that the third summand is also compact.

We write the fourth summand as

$$-M_\phi R_i^A \cdot \frac{(P^A)^{\frac{1}{2}}}{(1+P^A)^{\frac{1}{2}} \cdot ((P^A)^{\frac{1}{2}} + (1+P^A)^{\frac{1}{2}})} \cdot (P^A)^{-\frac{1}{2}} \Delta^{\frac{1}{2}} \cdot \Delta^{-\frac{1}{2}} M_\phi.$$

Since the functions $\{\phi \cdot a_{ij}^\Phi\}_{i,j=1}^{2d}$ are bounded, it follows from Lemma 3.2.5 that the first factor is bounded. The second factor is obviously bounded. The third factor is bounded by Lemma 3.2.5. The last factor belongs to $\mathcal{L}_{2d+2,\infty}$ by Cwikel estimate (see Theorem 1.1 in [48]).

This completes the proof. □

3.3.3 *Equivariant Behaviour of the Principal Symbol Under Global Diffeomorphisms*

Theorem 3.3.7 *Let $\Phi : \mathbb{H}^d \to \mathbb{H}^d$ be a diffeomorphism. Let $x \in \Pi$ be compactly supported (say, in $K \subset \mathbb{H}^d$). Suppose that Φ is Heisenberg in some neighborhood of K and is affine outside of some ball. We have*

$$U_\Phi^{-1} x U_\Phi \in \Pi,$$

$$\mathrm{sym}(U_\Phi^{-1} x U_\Phi) = \pi_{A^\Phi}(\mathrm{sym}(x)) \circ \Phi^{-1}.$$

The rest of this subsection is devoted to the proof of this theorem.

3.3 Invariance of Principal Symbol Map

Lemma 3.3.8 *Let* $(f_k)_{k=1}^m \subset \mathcal{A}_1$ *and* $(g_k)_{k=1}^m \subset \mathcal{A}_2$. *We have*

$$\prod_{k=1}^m \pi_1(f_k)\pi_2(g_k) \in \pi_1\Big(\prod_{k=1}^m f_k\Big)\pi_2\Big(\prod_{k=1}^m g_k\Big) + \mathcal{K}(L_2(\mathbb{R}^d)).$$

Proof We prove the assertion by induction on m. For $m = 1$, there is nothing to prove. So we only have to prove the step of induction.

Let us prove the assertion for $m = 2$. We have

$$\pi_1(f_1)\pi_2(g_1)\pi_1(f_2)\pi_2(g_2) = [\pi_2(g_1), \pi_1(f_1 f_2)] \cdot \pi_2(g_2)$$
$$+ [\pi_1(f_1), \pi_2(g_1)] \cdot \pi_1(f_2)\pi_2(g_2)$$
$$+ \pi_1(f_1 f_2)\pi_2(g_1 g_2).$$

By Lemma 2.2.1, we have

$$[\pi_1(f_1), \pi_2(g_1)], [\pi_2(g_1), \pi_1(f_1 f_2)] \in \mathcal{K}(L_2(\mathbb{H}^d)).$$

Therefore,

$$\pi_1(f_1)\pi_2(g_1)\pi_1(f_2)\pi_2(g_2) \in \pi_1(f_1 f_2)\pi_2(g_1 g_2) + \mathcal{K}(L_2(\mathbb{H}^d)).$$

This proves the assertion for $m = 2$.

It remains to prove the step of induction. Suppose the assertion holds for some $m \geq 2$ and let us prove it for $m + 1$. Clearly,

$$\prod_{k=1}^{m+1} \pi_1(f_k)\pi_2(g_k) = \pi_1(f_1)\pi_2(g_1) \cdot \prod_{k=2}^{m+1} \pi_1(f_k)\pi_2(g_k).$$

Using the inductive assumption, we obtain

$$\prod_{k=1}^{m+1} \pi_1(f_k)\pi_2(g_k) \in \pi_1(f_1)\pi_2(g_1) \cdot \pi_1\Big(\prod_{k=2}^{m+1} f_k\Big)\pi_2\Big(\prod_{k=2}^{m+1} g_k\Big) + \mathcal{K}(L_2(\mathbb{H}^d)).$$

Using the assertion for $m = 2$, we obtain

$$\pi_1(f_1)\pi_2(g_1) \cdot \pi_1\Big(\prod_{k=2}^{m+1} f_k\Big)\pi_2\Big(\prod_{k=2}^{m+1} g_k\Big) \in \pi_1\Big(\prod_{k=1}^{m+1} f_k\Big)\pi_2\Big(\prod_{k=1}^{m+1} g_k\Big) + \mathcal{K}(L_2(\mathbb{H}^d)).$$

Combining the last two equations, we obtain

$$\prod_{k=1}^{m+1} \pi_1(f_k)\pi_2(g_k) \in \pi_1(\prod_{k=1}^{m+1} f_k)\pi_2(\prod_{k=1}^{m+1} g_k) + \mathcal{K}(L_2(\mathbb{H}^d)).$$

This establishes the step of induction and, hence, completes the proof of the lemma. □

Lemma 3.3.9 *Let $\phi \in C_c(\mathbb{H}^d)$. Let $\Phi : \mathbb{H}^d \to \mathbb{H}^d$ be a diffeomorphism. Suppose that Φ is Heisenberg in some neighborhood of $\mathrm{supp}(\phi)$ and is affine outside of some ball. Suppose that $A : \mathbb{H}^d \to \mathrm{Aut}(\mathbb{H}^d)$ be a smooth mapping which is constant outside some ball and such that $A = A^\Phi$ near the support of ϕ. Suppose A and A^{-1} are bounded. We have*

$$U_\Phi^{-1} M_\phi R_i M_\phi U_\Phi, \; U_\Phi^{-1} M_\phi R_i^* M_\phi U_\Phi \in \Pi,$$

$$\mathrm{sym}(U_\Phi^{-1} M_\phi R_i M_\phi U_\Phi) = \pi_A(\mathrm{sym}(M_\phi R_i M_\phi)) \circ \Phi^{-1},$$

$$\mathrm{sym}(U_\Phi^{-1} M_\phi R_i^* M_\phi U_\Phi) = \pi_A(\mathrm{sym}(M_\phi R_i^* M_\phi)) \circ \Phi^{-1}.$$

Proof It is immediate that

$$U_\Phi^{-1} M_\phi R_i M_\phi U_\Phi = M_{\phi \circ \Phi^{-1}} U_\Phi^{-1} R_i U_\Phi M_{\phi \circ \Phi^{-1}}.$$

Applying Theorem 3.3.4 to the function $\phi \circ \Phi^{-1}$, diffeomorphism Φ and mapping $A \circ \Phi^{-1}$, we obtain

$$M_{\phi \circ \Phi^{-1}} U_\Phi^{-1} R_i U_\Phi M_{\phi \circ \Phi^{-1}} - M_{\phi \circ \Phi^{-1}} R_i^{A \circ \Phi^{-1}} M_{\phi \circ \Phi^{-1}} \in \mathcal{K}(L_2(\mathbb{H}^d)).$$

The assertion follows now by applying Theorems 3.2.2 and 1.3.2. □

Lemma 3.3.10 *Let $g \in C^*(\{R_k\}_{k=1}^{2d})$ and $f \in C_c(\mathbb{H}^d)$. Let Φ be a diffeomorphism. Suppose that Φ is Heisenberg in some neighborhood of $\mathrm{supp}(f)$ and affine outside of some ball. Suppose that $A : \mathbb{H}^d \to \mathrm{Aut}(\mathbb{H}^d)$ be a smooth mapping which is constant outside some ball and such that $A = A^\Phi$ in some neighborhood of $\mathrm{supp}(f)$ Suppose A and A^{-1} are bounded. We have*

$$U_\Phi^{-1} \pi_1(f) \pi_2(g) U_\Phi \in \Pi$$

and

$$\mathrm{sym}\left(U_\Phi^{-1} \pi_1(f)\pi_2(g) U_\Phi\right) = \pi_A(\mathrm{sym}(\pi_1(f)\pi_2(g))) \circ \Phi^{-1}.$$

3.3 Invariance of Principal Symbol Map

Proof Denote by $\mathrm{Alg}(\{R_i\}_{i=1}^{2d})$ the $*$-algebra generated by $\{R_i\}_{i=1}^{2d}$. Also, denote $R_i = R_{i-2d}^*$ for $2d+1 \le i \le 4d$.

Suppose first that $g \in \mathrm{Alg}(\{R_i\}_{i=1}^{2d})$ is monomial. Let $g = \prod_{l=1}^n R_{i(l)}$. Let $\phi \in C_c^\infty(\mathbb{H}^d)$ be such that $f \cdot \phi = f$ and $A = A^\Phi$ in some neighborhood of $\mathrm{supp}(\phi)$. Obviously,

$$\pi_1(f)\pi_2(g) = \pi_1(f \cdot \phi^{2n}) \cdot \pi_2(g) = \pi_1(f) \cdot \pi_1(\phi^{2n})\pi_2(g).$$

Set $f_l = \phi^2$, $1 \le l \le n$. We have

$$\pi_1(\phi^{2n})\pi_2(g) = \pi_1(\prod_{l=1}^n f_l)\pi_2(\prod_{l=1}^n R_{i(l)}).$$

By Lemma 3.3.8, we have

$$\pi_1(\phi^{2n})\pi_2(g) \in \prod_{l=1}^n \pi_1(f_l)\pi_2(R_{i(l)}) + \mathcal{K}(L_2(\mathbb{H}^d)).$$

Note that

$$\pi_1(f_l)\pi_2(R_{i(l)}) = \pi_1(\phi^2)\pi_2(R_{i(l)}) = \pi_1(\phi)\pi_2(R_{i(l)})\pi_1(\phi) + \pi_1(\phi) \cdot [\pi_1(\phi), \pi_2(R_{m(l)})].$$

By Lemma 2.2.1, we have

$$\pi_1(f_l)\pi_2(R_{i(l)}) \in \pi_1(\phi)\pi_2(R_{i(l)})\pi_1(\phi) + \mathcal{K}(L_2(\mathbb{H}^d)), \quad 1 \le l \le n.$$

Thus,

$$\pi_1(\phi^{2n})\pi_2(g) = \prod_{l=1}^n \pi_1(\phi)\pi_2(R_{i(l)})\pi_1(\phi) + Z, \quad Z \in \mathcal{K}(L_2(\mathbb{H}^d)).$$

Consequently,

$$U_\Phi^{-1}\pi_1(f)\pi_2(g)U_\Phi$$
$$= U_\Phi^{-1}\pi_1(f)U_\Phi \cdot \prod_{l=1}^n U_\Phi^{-1}\pi_1(\phi)\pi_2(R_{i(l)})\pi_1(\phi)U_\Phi + U_\Phi^{-1}\pi_1(f)ZU_\Phi.$$

It is immediate that $U_\Phi^{-1}\pi_1(f)U_\Phi \in \Pi$. By Lemma 3.3.9, we have

$$U_\Phi^{-1}\pi_1(\phi)\pi_2(R_{i(l)})\pi_1(\phi)U_\Phi \in \Pi, \quad 1 \le l \le n.$$

Since $Z \in \mathcal{K}(L_2(\mathbb{H}^d))$, it follows from Theorem 1.3.2 that

$$U_\Phi^{-1} \pi_1(f) Z U_\Phi \in \mathcal{K}(L_2(\mathbb{H}^d)) \subset \Pi.$$

Thus,

$$U_\Phi^{-1} \pi_1(f) \pi_2(g) U_\Phi \in \Pi.$$

Since sym is a $*$-homomorphism, it follows that

$$\mathrm{sym}(U_\Phi^{-1} \pi_1(f) \pi_2(g) U_\phi) = \mathrm{sym}(U_\Phi^{-1} \pi_1(f) Z U_\Phi) +$$
$$+ \mathrm{sym}(U_\Phi^{-1} \pi_1(f) U_\Phi)$$
$$\cdot \prod_{l=1}^{n} \mathrm{sym}(U_\Phi^{-1} \pi_1(\phi) \pi_2(R_{i(l)}) \pi_1(\phi) U_\Phi).$$

Since $U_\Phi^{-1} \pi_1(f) Z U_\Phi$ is compact, it follows that

$$\mathrm{sym}(U_\Phi^{-1} \pi_1(f) Z U_\Phi) = 0.$$

By Lemma 3.3.9, we have

$$\mathrm{sym}(U_\Phi^{-1} \pi_1(f) \pi_2(g) U_\phi)$$
$$= \left(\pi_A(\mathrm{sym}(\pi_1(f))) \cdot \prod_{l=1}^{n} \pi_A(\mathrm{sym}(\pi_1(\phi)\pi_2(R_{i(l)})\pi_1(\phi))) \right) \circ \Phi^{-1}.$$

Since π_A is a $*$-homomorphism, we have

$$\mathrm{sym}(U_\Phi^{-1} \pi_1(f) \pi_2(g) U_\phi)$$
$$= \pi_A \left(\mathrm{sym}(\pi_1(f)) \cdot \prod_{l=1}^{n} (\phi^2 \otimes R_{i(l)}) \mathrm{sym}(\pi_1(\phi)\pi_2(R_{i(l)})\pi_1(\phi)) \right) \circ \Phi^{-1}$$
$$= \pi_A \left((f \otimes 1) \cdot \prod_{l=1}^{n} (\phi^2 \otimes R_{i(l)}) \right) \circ \Phi^{-1}$$
$$= \pi_A \left(f \otimes \prod_{l=1}^{n} R_{i(l)} \right) \circ \Phi^{-1}$$
$$= \pi_A \left(\mathrm{sym}(\pi_1(f)\pi_2(g)) \right) \circ \Phi^{-1}.$$

This proves the assertion for the case when $g \in \mathrm{Alg}(\{R_i\}_{i=1}^{2d})$ is monomial.

3.3 Invariance of Principal Symbol Map

By linearity, the same assertion holds if $g \in \mathrm{Alg}(\{R_i\}_{i=1}^{2d})$. To prove the assertion in general, let $g \in C^*(\{R_i\}_{i=1}^{2d})$ and consider a sequence $\{g_n\}_{n\geq 1} \subset \mathrm{Alg}(\{R_i\}_{i=1}^{2d})$ such that $g_n \to g$ in the uniform norm. We have

$$U_\Phi^{-1}\pi_1(f)\pi_2(g_n)U_\Phi \to U_\Phi^{-1}\pi_1(f)\pi_2(g)U_\Phi$$

in the uniform norm. Noting that

$$U_\Phi^{-1}\pi_1(f)\pi_2(g_n)U_\Phi \in \Pi, \quad n \geq 1,$$

we obtain

$$U_\Phi^{-1}\pi_1(f)\pi_2(g)U_\Phi \in \Pi$$

and

$$\begin{aligned}\mathrm{sym}(U_\Phi^{-1}\pi_1(f)\pi_2(g)U_\Phi) &= \lim_{n\to\infty} \mathrm{sym}(U_\Phi^{-1}\pi_1(f)\pi_2(g_n)U_\Phi) \\ &= \lim_{n\to\infty} \pi_A(\mathrm{sym}(\pi_1(f)\pi_2(g_n)) \circ \Phi^{-1}) \\ &= \pi_A(\mathrm{sym}(\pi_1(f)\pi_2(g)) \circ \Phi^{-1}).\end{aligned}$$

□

Proof of Theorem 3.3.7 Let $A : \mathbb{H}^d \to \mathrm{Aut}(\mathbb{H}^d)$ be a smooth mapping constant outside of some ball such that $A = A^\Phi$ in some neighborhood of K. Suppose also that A and A^{-1} are bounded.

Choose $\phi \in C_c^\infty(\mathbb{H}^d)$ such that $\phi = 1$ in some neighborhood of K and such that Φ is Heisenberg and $A = A^\Phi$ on $\mathrm{supp}(\phi)$.

By the definition of the C^*-algebra Π, there exists a sequence $(x_n)_{n\geq 1}$ in the $*$-algebra generated by $\pi_1(\mathcal{A}_1)$ and $\pi_2(\mathcal{A}_2)$ such that $x_n \to x$ in the uniform norm.

Choose $\psi \in C_c^\infty(\mathbb{H}^d)$ such that $\phi\psi = \psi$ and such that $\psi = 1$ on K. Clearly, $x = \pi_1(\psi)x$. Replacing x_n with $\pi_1(\psi)x_n$ if needed, we may assume without loss of generality that $x_n = \pi_1(\phi)x_n$.

We can write

$$x_n = \sum_{l=1}^{l_n}\prod_{k=1}^{k_n} \pi_1(f_{n,k,l})\pi_2(g_{n,k,l}).$$

By Lemma 3.3.8, we have

$$x_n \in \sum_{l=1}^{l_n} \pi_1\left(\prod_{k=1}^{k_n} f_{n,k,l}\right)\pi_2\left(\prod_{k=1}^{k_n} g_{n,k,l}\right) + \mathcal{K}(L_2(\mathbb{H}^d)).$$

Denote for brevity,

$$f_{n,l} = \prod_{k=1}^{k_n} f_{n,k,l} \in \mathcal{A}_1, \quad g_{n,l} = \prod_{k=1}^{k_n} g_{n,k,l} \in \mathcal{A}_2.$$

We have

$$x_n = y_n + z_n, \quad y_n = \sum_{l=1}^{l_n} \pi_1(f_{n,l})\pi_2(g_{n,l}), \quad z_n \in \mathcal{K}(L_2(\mathbb{H}^d)).$$

Replacing y_n with $\pi_1(\phi)y_n$ and z_n with $\pi_1(\phi)z_n$ if needed, we may assume without loss of generality that each $f_{n,l}$ is supported on $\mathrm{supp}(\phi)$.

By Lemma 3.3.10, we have

$$U_\Phi^{-1} y_n U_\Phi \in \Pi$$

and

$$\mathrm{sym}(U_\Phi^{-1} y_n U_\Phi) = \pi_A(\mathrm{sym}(y_n)) \circ \Phi^{-1}.$$

By Theorem 1.3.2, we have $U_\Phi^{-1} z_n U_\Phi \in \Pi$ and

$$\mathrm{sym}(U_\Phi^{-1} z_n U_\Phi) = 0.$$

Thus, $U_\Phi^{-1} x_n U_\Phi \in \Pi$ and

$$\mathrm{sym}(U_\Phi^{-1} x_n U_\Phi) = \pi_A(\mathrm{sym}(y_n)) \circ \Phi^{-1} = \pi_A(\mathrm{sym}(x_n)) \circ \Phi^{-1}.$$

Since Π is a C^*-algebra and $U_\Phi^{-1} x_n U_\Phi \to U_\Phi^{-1} x U_\Phi$ in the uniform norm, it follows that $U_\Phi^{-1} x U_\Phi \in \Pi$ and

$$\mathrm{sym}(U_\Phi^{-1} x_n U_\Phi) \to \mathrm{sym}(U_\Phi^{-1} x U_\Phi)$$

in the uniform norm. In other words,

$$\pi_A(\mathrm{sym}(x_n)) \circ \Phi^{-1} \to \mathrm{sym}(U_\Phi^{-1} x U_\Phi)$$

in the uniform norm. Since $\mathrm{sym}(x_n) \to \mathrm{sym}(x)$ in the uniform norm, it follows that

$$\mathrm{sym}(U_\Phi^{-1} x U_\Phi) = \pi_A(\mathrm{sym}(x)) \circ \Phi^{-1}.$$

Since sym(x) is supported on K and $A = A^\Phi$ in the neighborhood of K, it follows that one can write $\pi_A(\text{sym}(x)) = \pi_{A^\Phi}(\text{sym}(x))$. □

3.3.4 Equivariant Behaviour of the Principal Symbol Under Local Diffeomorphisms

Notation 3.3.11 Let H be a Hilbert space and let $p \in B(H)$ be a projection.

(i) If $S \in B(H)$ is such that $S = pSp$, then we define the operator $\text{Rest}_p(S) \in B(pH)$ by setting $\text{Rest}_p(S) = S|_{pH}$.
(ii) If $S \in B(pH)$, then we define $\text{Ext}_p(S) \in B(H)$ by setting $\text{Ext}_p(S) = S \circ p$.

Notation 3.3.12 Let $\Omega \subset \mathbb{H}^d$. If $S \in B(L_2(\mathbb{H}^d))$ is such that $S = M_{\chi_\Omega} S M_{\chi_\Omega}$, then $\text{Rest}_\Omega(S) \in B(L_2(\Omega))$ is a shorthand for $\text{Rest}_{M_{\chi_\Omega}}(S)$. If $S \in B(L_2(\Omega))$, then $\text{Ext}_\Omega(S) \in B(L_2(\mathbb{H}^d))$ is a shorthand for $\text{Ext}_{M_{\chi_\Omega}}(S)$.

Theorem 3.3.13 *Let $\Omega, \Omega' \subset \mathbb{H}^d$ be open sets and let $\Phi : \Omega \to \Omega'$ be a Heisenberg diffeomorphism. Let $x \in \Pi$ be compactly supported in Ω. We have*

$$\text{Ext}_{\Omega'}\left(U_\Phi^{-1} \cdot \text{Rest}_\Omega(x) \cdot U_\Phi\right) \in \Pi,$$

$$\text{sym}\left(\text{Ext}_{\Omega'}\left(U_\Phi^{-1} \cdot \text{Rest}_\Omega(x) \cdot U_\Phi\right)\right) = \pi_{A^\Phi}(\text{sym}(x)) \circ \Phi^{-1}.$$

Theorem 3.3.13 is supposed to be a corollary of Theorem 3.3.7. To demonstrate this is indeed the case, we need the following results from [43].

In this subsection, $B(p, r)$ denotes the (Euclidean) open ball with radius r centered at p.

Lemma 3.3.14 *Let $\Omega \subset \mathbb{H}^d$ be an open set and let $\Phi : \Omega \to \mathbb{H}^d$ be a smooth mapping. If $p \in \Omega$ is such that $\det(J_\Phi(p)) \neq 0$, then there exists a diffeomorphism $\Phi_p : \mathbb{H}^d \to \mathbb{H}^d$ such that*

(i) $\Phi_p = \Phi$ on $B(p, r_1(p))$ with some $r_1(p) > 0$;
(ii) Φ_p is affine outside $B(p, r_2(p))$ for some $r_2(p) < \infty$;

Proof Write $\mathbb{H}^d = \mathbb{R}^{2d+1}$. Apply Lemma 6.2 in [43] with $2d + 1$ instead of d. □

Lemma 3.3.15 *Let $\Omega, \Omega' \subset \mathbb{H}^d$ and let $\Phi : \Omega \to \Omega'$ be a diffeomorphism. Let $B \subset \Omega$ be a ball and let $\Phi_0 : \mathbb{H}^d \to \mathbb{H}^d$ be a diffeomorphism such that $\Phi_0 = \Phi$ on B. If $x \in B(L_2(\mathbb{H}^d))$ is supported on B, then*

$$\text{Ext}_{\Omega'}\left(U_\Phi^{-1} \cdot \text{Rest}_\Omega(x) \cdot U_\Phi\right) = U_{\Phi_0}^{-1} x U_{\Phi_0}.$$

Proof Write $\mathbb{H}^d = \mathbb{R}^{2d+1}$. Apply Lemma 6.2 in [43] with $2d + 1$ instead of d. □

Proof of Theorem 3.3.13 Let the operator x be supported on a compact set $K \subset \Omega$. Let $p \in K$. Let a diffeomorphism $\Phi_p : \mathbb{H}^d \to \mathbb{H}^d$ and numbers $r_1(p)$ and $r_2(p)$ be as in Lemma 3.3.14. Without loss of generality, $r_1(p)$ is so small that $B(p, r_1(p)) \subset \Omega$. Since Φ is Heisenberg on Ω and $\Phi_p = \Phi$ on $B(p, r_1(p))$, it follows that Φ_p is Heisenberg on $B(p, r_1(p))$.

The collection $\{B(p, r_1(p))\}_{p \in K}$ is an open cover of K. By compactness, one can choose a finite sub-cover $\{B(p_n, r_1(p_n))\}_{n=1}^{N}$.

Let $\{\phi_n\}_{n=1}^{N}$ be such that $\phi_n \in C_c^{\infty}(B(p_n, r_1(p_n)))$ and

$$\sum_{n=1}^{N} \phi_n^2 = 1 \quad \text{on } K.$$

Set

$$x_0 = \sum_{m=1}^{N} M_{\phi_m}[M_{\phi_m}, x], \quad x_n = M_{\phi_n} x M_{\phi_n}, \quad 1 \leq n \leq N.$$

We write

$$x = \sum_{n=1}^{N} M_{\phi_n^2} x = \sum_{n=0}^{N} x_n. \tag{3.3.1}$$

If $1 \leq n \leq N$, then x_n is supported on the ball $B(p_n, r_1(p_n))$. By Lemma 3.3.15, we have

$$\mathrm{Ext}_{\Omega'}\left(U_{\Phi}^{-1} \cdot \mathrm{Rest}_{\Omega}(x_n) \cdot U_{\Phi}\right) = U_{\Phi_{p_n}}^{-1} x_n U_{\Phi_{p_n}}.$$

By construction, x_n is supported in $B(p_n, r_1(p_n))$ and Φ_{p_n} is Heisenberg in $B(p_n, r_1(p_n))$ and is affine outside some ball. By Theorem 3.3.7, we have $U_{\Phi_{p_n}}^{-1} x_n U_{\Phi_{p_n}} \in \Pi$ and

$$\mathrm{sym}(U_{\Phi_{p_n}}^{-1} x_n U_{\Phi_{p_n}}) = \pi_{A\Phi_{p_n}}(\mathrm{sym}(x_n)) \circ \Phi_{p_n}^{-1}.$$

Since $\mathrm{sym}(x_n)$ is supported in $B(p_n, r_1(p_n))$ and $\Phi_{p_n} = \Phi$ in $B(p_n, r_1(p_n))$, it follows that

$$\pi_{A\Phi_{p_n}}(\mathrm{sym}(x_n)) \circ \Phi_{p_n}^{-1} = \pi_{A\Phi}(\mathrm{sym}(x_n)) \circ \Phi^{-1}.$$

Thus,

$$\mathrm{Ext}_{\Omega'}\left(U_{\Phi}^{-1} \cdot \mathrm{Rest}_{\Omega}(x_n) \cdot U_{\Phi}\right) \in \Pi \tag{3.3.2}$$

3.3 Invariance of Principal Symbol Map

and

$$\mathrm{sym}\Big(\mathrm{Ext}_{\Omega'}\Big(U_\Phi^{-1} \cdot \mathrm{Rest}_\Omega(x_n) \cdot U_\Phi\Big)\Big) = \pi_{A^\Phi}(\mathrm{sym}(x_n)) \circ \Phi^{-1}. \tag{3.3.3}$$

If $n = 0$, then, by Lemma 2.2.1, x_0 is compact. Clearly, x_0 is compactly supported in Ω. If $A \in B(L_2(\Omega'))$ is compact, then $\mathrm{Ext}_{\Omega'}(A) \in B(L_2(\mathbb{H}^d))$ is also compact. Therefore,

$$\mathrm{Ext}_{\Omega'}\Big(U_\Phi^{-1} \cdot \mathrm{Rest}_\Omega(x_0) \cdot U_\Phi\Big) \in \mathcal{K}(L_2(\mathbb{H}^d)) \subset \Pi \tag{3.3.4}$$

and

$$\mathrm{sym}\Big(\mathrm{Ext}_{\Omega'}\Big(U_\Phi^{-1} \cdot \mathrm{Rest}_\Omega(x_0) \cdot U_\Phi\Big)\Big) = 0. \tag{3.3.5}$$

Combining (3.3.1), (3.3.2) and (3.3.4), we obtain

$$\mathrm{Ext}_{\Omega'}\Big(U_\Phi^{-1} \cdot \mathrm{Rest}_\Omega(x) \cdot U_\Phi\Big) = \sum_{n=0}^{N} \mathrm{Ext}_{\Omega'}\Big(U_\Phi^{-1} \cdot \mathrm{Rest}_\Omega(x_n) \cdot U_\Phi\Big) \in \Pi.$$

Now, combining (3.3.1), (3.3.3) and (3.3.5), we obtain

$$\mathrm{sym}\Big(\mathrm{Ext}_{\Omega'}\Big(U_\Phi^{-1} \cdot \mathrm{Rest}_\Omega(x) \cdot U_\Phi\Big)\Big) = \sum_{n=0}^{N} \mathrm{sym}\Big(\mathrm{Ext}_{\Omega'}\Big(U_\Phi^{-1} \cdot \mathrm{Rest}_\Omega(x_n) \cdot U_\Phi\Big)\Big)$$

$$= \sum_{n=1}^{N} \pi_{A^\Phi}(\mathrm{sym}(x_n)) \circ \Phi^{-1}$$

$$= \pi_{A^\Phi}(\mathrm{sym}(x)) \circ \Phi^{-1}$$

$$\quad - \pi_{A^\Phi}(\mathrm{sym}(x_0)) \circ \Phi^{-1}$$

$$= \pi_{A^\Phi}(\mathrm{sym}(x)) \circ \Phi^{-1}.$$

$\square$

Chapter 4
Principal Symbol on Contact Manifolds

4.1 Principal Symbol on Compact Contact Manifolds

4.1.1 Globalisation Theorem

In this short subsection, we remind the reader the globalisation theorem established in [43].

Definition 4.1.1 Let X be a compact manifold with an atlas $\{(\mathcal{U}_i, h_i)\}_{i \in \mathbb{I}}$. Let $\mathfrak{B}$ be the Borel σ-algebra on X and let ν be a countably additive measure on $\mathfrak{B}$. We say that $\{\mathcal{A}_i\}_{i \in \mathbb{I}}$ are local algebras if

(i) for every $i \in \mathbb{I}$, $\mathcal{A}_i$ is a $*$-subalgebra in $B(L_2(X, \nu))$;
(ii) for every $i \in \mathbb{I}$, elements of $\mathcal{A}_i$ are compactly supported in $\mathcal{U}_i$;
(iii) for every $i, j \in \mathbb{I}$, if $S \in \mathcal{A}_i$ is compactly supported in $\mathcal{U}_i \cap \mathcal{U}_j$, then $S \in \mathcal{A}_j$;
(iv) for every $i \in \mathbb{I}$, if $S \in \mathcal{K}(L_2(X, \nu))$ is compactly supported in $\mathcal{U}_i$, then $S \in \mathcal{A}_i$;
(v) for every $i \in \mathbb{I}$, if $\phi \in C_c(\mathcal{U}_i)$, then $M_\phi \in \mathcal{A}_i$;
(vi) for every $i \in \mathbb{I}$, if $\phi \in C_c(\mathcal{U}_i)$, then the closure of $M_\phi \mathcal{A}_i M_\phi$ in the uniform norm is contained in $\mathcal{A}_i$;
(vii) for every $i \in \mathbb{I}$, if $S \in \mathcal{A}_i$ and if $\phi \in C_c(\mathcal{U}_i)$, then $[S, M_\phi] \in \mathcal{K}(L_2(X, \nu))$.

Definition 4.1.2 In the setting of Definition 4.1.1, we say that $S \in \mathcal{A}$ if

(i) for every $i \in \mathbb{I}$ and for every $\phi \in C_c(\mathcal{U}_i)$, we have $M_\phi S M_\phi \in \mathcal{A}_i$;
(ii) for every $\psi \in C(X)$, the commutator $[S, M_\psi]$ is compact.

Definition 4.1.3 Let $\mathcal{B}$ be a $*$-algebra. In the setting of Definition 4.1.1, $\{\hom_i\}_{i \in \mathbb{I}}$ are called local homomorphisms if

(i) for every $i \in \mathbb{I}$, $\hom_i : \mathcal{A}_i \to \mathcal{B}$ is a $*$-homomorphism;
(ii) for every $i, j \in \mathbb{I}$, we have $\hom_i = \hom_j$ on $\mathcal{A}_i \cap \mathcal{A}_j$;
(iii) $S \in \mathcal{A}_i$ is compact iff $\hom_i(S) = 0$;

(iv) there exists a *-homomorphism Hom : $C(X) \to \mathcal{B}$ such that

$$\hom_i(M_\phi) = \text{Hom}(\phi), \quad \phi \in C_c(\mathcal{U}_i), \quad i \in \mathbb{I}.$$

Theorem 4.1.4 *In the setting of Definitions 4.1.1, 4.1.3 and 4.1.2, we have*

(i) $\mathcal{A}$ *is a unital C^*-subalgebra in $B(L_2(X, \nu))$ which contains $\mathcal{A}_i$ for every $i \in \mathbb{I}$ and $\mathcal{K}(L_2(X, \nu))$;*
(ii) *there exists a unique *-homomorphism* hom : $\mathcal{A} \to \mathcal{B}$ *such that*

    (a) hom $= \hom_i$ on $\mathcal{A}_i$ for every $i \in \mathbb{I}$;
    (b) $\ker(\hom) = \mathcal{K}(L_2(X, \nu))$.

4.1.2 Construction of the Principal Symbol Mapping

Let X be a compact contact manifold with a Heisenberg atlas $\{(\mathcal{U}_i, h_i)\}_{i \in \mathbb{I}}$. Let $\mathfrak{B}$ be the Borel σ-algebra on the manifold X and let $\nu : \mathfrak{B} \to \mathbb{R}$ be a continuous positive density. It is immediate that the mapping $h_i : (\mathcal{U}_i, \nu) \to (\Omega_i, \nu \circ h_i^{-1})$ preserves the measure. Define an isometry $W_i : L_2(\mathcal{U}_i, \nu) \to L_2(\Omega_i, \nu \circ h_i^{-1})$ by setting

$$W_i f = f \circ h_i^{-1}, \quad f \in L_2(\mathcal{U}_i, \nu).$$

If T is compactly supported in $\mathcal{U}_i$, then $W_i T W_i^{-1}$ is understood as an element of the algebra $B(L_2(\Omega_i, \nu \circ h_i^{-1}))$. The latter operator is compactly supported in Ω_i. By Definition 1.3.7, exactly the same operator also belongs to $B(L_2(\Omega_i))$ and, therefore, can be extended to an element $\text{Ext}_{\Omega_i}(W_i T W_i^{-1})$ of $B(L_2(\mathbb{H}^d))$. For notation Π below we refer to Definition 1.3.1.

Definition 4.1.5 Let X be a compact contact manifold with a Heisenberg atlas $\{(\mathcal{U}_i, h_i)\}_{i \in \mathbb{I}}$ and let ν be a continuous positive density on X. For every $i \in \mathbb{I}$, let Π_i consist of operators $T \in B(L_2(X, \nu))$ compactly supported in $\mathcal{U}_i$ such that

$$\text{Ext}_{\Omega_i}(W_i T W_i^{-1}) \in \Pi.$$

For example, every operator M_ϕ, $\phi \in C_c(\mathcal{U}_i)$ belongs to Π_i.

Next, we check that the codomain of the principal symbol mapping is well defined. Recall that, for any $i, j \in \mathbb{I}$ such that $\mathcal{U}_i \cap \mathcal{U}_j \neq \emptyset$, the diffeomorphism

$$\Phi_{i,j} : \Omega_{i,j} = h_i(\mathcal{U}_i \cap \mathcal{U}_j) \subset \mathbb{R}^{2d+1} \to \Omega_{j,i} = h_j(\mathcal{U}_i \cap \mathcal{U}_j) \subset \mathbb{R}^{2d+1}$$

introduced in Notation 1.3.4 is a Heisenberg diffeomorphism.

4.1 Principal Symbol on Compact Contact Manifolds

Lemma 4.1.6 *The family $\varpi = \{\pi_{i,j}\}_{i,j \in \mathbb{I}}$ of continuous mappings $\pi_{i,j} : \mathcal{U}_i \cap \mathcal{U}_j \to \mathrm{Aut}(C^*(\{R_k\}_{k=1}^{2d})$ defined by setting*

$$\pi_{i,j} = \pi_{A^{\Phi_{i,j}} \circ h_i}, \quad i,j \in \mathbb{I} \text{ such that } \mathcal{U}_i \cap \mathcal{U}_j \neq \emptyset,$$

satisfies (1.3.3) *in Definition* 1.3.9.

Proof Let $i, j \in \mathbb{I}$ be such that $\mathcal{U}_i \cap \mathcal{U}_j \neq \emptyset$. It follows from (1.3.6) that

$$A^{\Phi_{j,i}}(\Phi_{i,j}(p')) \cdot A^{\Phi_{i,j}}(p') = 1_{\mathrm{Aut}(\mathbb{H}^d)}, \quad p' \in \Omega_{i,j}.$$

Using (1.3.8), we infer that

$$\pi_{A^{\Phi_{i,j}}(p')} \circ \pi_{A^{\Phi_{j,i}}(\Phi_{i,j}(p'))} = \mathrm{id}, \quad p' \in \Omega_{i,j}.$$

Setting $p' = h_i(p)$, $p \in \mathcal{U}_i \cap \mathcal{U}_j$, we arrive at

$$\pi_{i,j}(p) \circ \pi_{j,i}(p) = \pi_{A^{\Phi_{i,j}}(h_i(p))} \circ \pi_{A^{\Phi_{j,i}}(h_j(p))} = \mathrm{id}, \quad p \in \mathcal{U}_i \cap \mathcal{U}_j.$$

This proves the first equality in (1.3.3).

Let $i, j, k \in \mathbb{I}$ be such that $\mathcal{U}_i \cap \mathcal{U}_j \cap \mathcal{U}_k \neq \emptyset$. It follows from (1.3.6) that

$$A^{\Phi_{j,k}}(\Phi_{i,j}(p')) \cdot A^{\Phi_{i,j}}(p') = A^{\Phi_{i,k}}(p'), \quad p' \in \Omega_{i,j} \cap \Omega_{i,k}.$$

Again using (1.3.6), we write

$$A^{\Phi_{k,i}}(\Phi_{i,k}(p')) \cdot A^{\Phi_{j,k}}(\Phi_{i,j}(p')) \cdot A^{\Phi_{i,j}}(p') = A^{\Phi_{k,i}}(\Phi_{i,k}(p')) \cdot A^{\Phi_{i,k}}(p')$$
$$= 1_{\mathrm{Aut}(\mathbb{H}^d)}, \quad p' \in \Omega_{i,j} \cap \Omega_{i,k}.$$

Using (1.3.8), we infer that

$$\pi_{A^{\Phi_{i,j}}(p')} \circ \pi_{A^{\Phi_{j,k}}(\Phi_{i,j}(p'))} \circ \pi_{A^{\Phi_{k,i}}(\Phi_{i,k}(p'))} = \mathrm{id}, \quad p' \in \Omega_{i,j} \cap \Omega_{i,k}.$$

Setting $p' = h_i(p)$, $p \in \mathcal{U}_i \cap \mathcal{U}_j \cap \mathcal{U}_k$, we arrive at

$$\pi_{i,j}(p) \circ \pi_{j,k}(p) \circ \pi_{k,i}(p) = \pi_{A^{\Phi_{i,j}}(h_i(p))} \circ \pi_{A^{\Phi_{j,k}}(h_j(p))} \circ \pi_{A^{\Phi_{k,i}}(h_k(p))}$$
$$= \mathrm{id}, \quad p \in \mathcal{U}_i \cap \mathcal{U}_j \cap \mathcal{U}_k.$$

This proves the second equality in (1.3.3). □

Consider the triple $E_{\mathrm{hom}} = (X, C^*(\{R_k\}_{k=1}^{2d}), \varpi)$, where ϖ is given in Lemma 4.1.6. By Lemma 4.1.6, E_{hom} is a bundle of C^*-algebras in the sense of

Definition 1.3.9. Let $C(E_{\text{hom}})$ be the algebra of continuous sections of the C^*-algebra bundle $E_{\text{hom}} = (X, \mathcal{M}, \varpi)$ (see Definition 1.3.10).

For the notion sym below, we refer to Theorem 1.3.2. The mapping Ξ_i used in the Definition 4.1.7 is defined in Sect. 1.3, right after Theorem 1.3.13.

Definition 4.1.7 Let X be a compact contact manifold with a Heisenberg atlas $\{(\mathcal{U}_i, h_i)\}_{i \in \mathbb{I}}$. and let ν be a continuous positive density on X. For every $i \in \mathbb{I}$, the mapping $\text{sym}_i : \Pi_i \to C(E_{\text{hom}})$ is defined by the formula

$$\text{sym}_i(S) = \Xi_i \left(\text{sym}(\text{Ext}_{\Omega_i}(W_i S W_i^{-1})) \circ h_i \right), \quad S \in \Pi_i.$$

Theorem 4.1.8 *Let X be a compact contact manifold with a Heisenberg atlas $\{(\mathcal{U}_i, h_i)\}_{i \in \mathbb{I}}$. and let ν be a continuous positive density on X.*

(i) *Collection $\{\Pi_i\}_{i \in \mathbb{I}}$ introduced in Definition 4.1.5 satisfies all the conditions in Definition 4.1.1;*
(ii) *Collection $\{\text{sym}_i\}_{i \in \mathbb{I}}$ introduced in Definition 4.1.7 satisfies all the conditions in Definition 4.1.3.*

That is, the collection $\{\Pi_i\}_{i \in \mathbb{I}}$ of $*$-algebras and the collection $\{\text{sym}_i\}_{i \in \mathbb{I}}$ of $*$-homomorphisms satisfy the conditions in Theorem 4.1.4.

Definition 4.1.9 below is the culmination of the paper. Having this definition at hands, we easily prove Theorem 1.3.18.

Definition 4.1.9 Let X be a compact contact manifold with a Heisenberg atlas $\{(\mathcal{U}_i, h_i)\}_{i \in \mathbb{I}}$. and let ν be a continuous positive density on X.

(i) The domain Π_X of the principal symbol mapping is the C^*-algebra constructed in Theorem 4.1.4 from the collection $\{\Pi_i\}_{i \in \mathbb{I}}$.
(ii) The principal symbol mapping $\text{sym}_X : \Pi_X \to C(E_{\text{hom}})$ is the $*$-homomorphism constructed in Theorem 4.1.4 from the collection $\{\text{sym}_i\}_{i \in \mathbb{I}}$.

4.1.3 Proof of Theorem 4.1.8

Lemma 4.1.10 below verifies the condition (i) in Definitions 4.1.1 and 4.1.3.

Lemma 4.1.10 *For every $i \in \mathbb{I}$, we have*

(i) Π_i *is a $*$-subalgebra in $B(L_2(X, \nu))$;*
(ii) $\text{sym}_i : \Pi_i \to C(E_{\text{hom}})$ *is a $*$-homomorphism.*

Proof It is immediate that Π_i is a subalgebra in $B(L_2(X, \nu))$ and $\text{sym}_i : \Pi_i \to C(E_{\text{hom}})$ is a homomorphism. We need to show that Π_i is closed with respect to taking adjoints and sym_i is invariant with respect to this operation.

Given $S \in \Pi_i$, let us show that $S^* \in \Pi_i$. Recall that, due to Condition 1.3.7, $\nu \circ h_i^{-1}$ is absolutely continuous and its density denoted by a_i as well as its inverse

4.1 Principal Symbol on Compact Contact Manifolds

a_i^{-1} are assumed to be continuous in Ω_i. The following equality[1] is easy to verify directly:

$$W_i S^* W_i^{-1} = M_{a_i^{-1}} \cdot (W_i S W_i^{-1})^* \cdot M_{a_i}. \qquad (4.1.1)$$

However, by Definition 4.1.5, the operator S is compactly supported in $\mathcal{U}_i$. Hence, the operator $(W_i S W_i^{-1})^*$ is compactly supported in Ω_i. Choose $\phi \in C_c(\Omega_i)$ such that

$$(W_i S W_i^{-1})^* = M_\phi \cdot (W_i S W_i^{-1})^* = (W_i S W_i^{-1})^* \cdot M_\phi.$$

Thus,

$$W_i S^* W_i^{-1} = M_{a_i^{-1}\phi} \cdot (W_i S W_i^{-1})^* \cdot M_{a_i\phi}.$$

Thus,

$$\operatorname{Ext}_{\Omega_i}(W_i S^* W_i^{-1}) = M_{a_i^{-1}\phi} \cdot (\operatorname{Ext}_{\Omega_i}(W_i S W_i^{-1}))^* \cdot M_{a_i\phi}. \qquad (4.1.2)$$

Since $a_i\phi, a_i^{-1}\phi \in C_c(\mathbb{R}^d)$, it follows that every factor in the right hand side of (4.1.2) belongs to Π. Hence, so is the expression on the left hand side. In other words, $T^* \in \Pi_i$. Thus, Π_i is closed with respect to taking adjoints.

Recall that (by Theorem 1.3.2) sym is a $*$-homomorphism. Applying sym to the equality (4.1.2), we obtain

$$\operatorname{sym}(\operatorname{Ext}_{\Omega_i}(W_i S^* W_i^{-1})) = \operatorname{sym}(M_{a_i^{-1}\phi}) \cdot \operatorname{sym}((\operatorname{Ext}_{\Omega_i}(W_i S W_i^{-1}))^*) \cdot \operatorname{sym}(M_{a_i\phi})$$

$$= \operatorname{sym}(M_{\phi^2}) \cdot \operatorname{sym}(\operatorname{Ext}_{\Omega_i}(W_i S W_i^{-1}))^*$$

$$= \operatorname{sym}(\operatorname{Ext}_{\Omega_i}(M_{\phi^2} \cdot W_i S W_i^{-1}))^*.$$

It is clear that

$$M_{\phi^2} \cdot W_i S W_i^{-1} = W_i S W_i^{-1}.$$

---

[1] The operators M_{a_i} and $M_{a_i^{-1}}$ are unbounded. The equality should be understood as $LHS\xi = RHS\xi$ for every compactly supported $\xi \in L_2(\Omega_i)$.

Indeed, for such ξ, we have $\xi_1 = M_{a_i}\xi \in L_2(\Omega_i)$. Since S is compactly supported in $\mathcal{U}_i$, $(W_i S W_i^{-1})^*$ is compactly supported in Ω_i. It follows that the function $\xi_2 = (W_i S W_i^{-1})^*\xi_1$ is compactly supported in Ω_i. Hence, the function $M_{a_i^{-1}}\xi_2$ belongs to $L_2(\Omega_i)$ and the right hand side of (4.1.1) makes sense.

Thus,
$$\operatorname{sym}(\operatorname{Ext}_{\Omega_i}(W_i S^* W_i^{-1})) = \operatorname{sym}(\operatorname{Ext}_{\Omega_i}(W_i S W_i^{-1}))^*.$$

Since Ξ_i is an $*$-isomorphism, it follows from Definition 4.1.7 that
$$\operatorname{sym}_i(S^*) = \Xi_i\left(\operatorname{sym}(\operatorname{Ext}_{\Omega_i}(W_i S^* W_i^{-1})) \circ h_i\right)$$
$$= \left(\Xi_i\left(\operatorname{sym}(\operatorname{Ext}_{\Omega_i}(W_i S W_i^{-1})) \circ h_i\right)\right)^*$$
$$= \operatorname{sym}_i(S)^*.$$

□

The following lemma verifies the condition (iii) in Definition 4.1.1 and the condition (ii) in Definition 4.1.3.

Lemma 4.1.11 *Let $(\mathcal{U}_i, h_i)$ and $(\mathcal{U}_j, h_j)$ be charts. Let $S \in B(L_2(X, \nu))$ be compactly supported in $\mathcal{U}_i \cap \mathcal{U}_j$. If $S \in \Pi_i$, then $S \in \Pi_j$ and $\operatorname{sym}_i(S) = \operatorname{sym}_j(S)$.*

Proof Denote for brevity
$$S_i = \operatorname{Ext}_{\Omega_i}(W_i S W_i^{-1}), \quad S_j = \operatorname{Ext}_{\Omega_j}(W_j S W_j^{-1}).$$

Recall the notation $V_\Phi \xi = \xi \circ \Phi$ (provided that the image of the mapping Φ is contained in the domain of the function ξ). Since $W_j = V_{\Phi_{i,j}}^{-1} W_i$, it follows that
$$\operatorname{Rest}_{\Omega_j}(S_j) = U_{\Phi_{i,j}}^{-1} \cdot M_{|J_{\Phi_{i,j}}|^{\frac{1}{2}}} \cdot \operatorname{Rest}_{\Omega_i}(S_i) \cdot M_{|J_{\Phi_{i,j}}|^{-\frac{1}{2}}} \cdot U_{\Phi_{i,j}}. \quad (4.1.3)$$

Since S is compactly supported in $\mathcal{U}_i \cap \mathcal{U}_j$, it follows that S_i is compactly supported in $\Omega_{i,j}$. Let $A \subset \Omega_{i,j}$ be compact and such that
$$M_{\chi_A} \cdot S_i \cdot M_{\chi_A} = W_i S W_i^{-1}.$$

Using Tietze extension theorem, choose $\phi \in C(\mathbb{H}^d)$ such that $\phi^{-1} \in C(\mathbb{H}^d)$ and $\phi = |J_{\Phi_{i,j}}|^{\frac{1}{2}}$ on A. It follows that
$$\operatorname{Ext}_{\Omega_i}\left(M_{|J_{\Phi_{i,j}}|^{\frac{1}{2}}} \cdot \operatorname{Rest}_{\Omega_i}(S_i) \cdot M_{|J_{\Phi_{i,j}}|^{-\frac{1}{2}}}\right) = M_\phi S_i M_{\phi^{-1}} = S_i + [M_\phi, S_i] \cdot M_{\phi^{-1}}.$$

It follows from Lemma 2.2.1 that
$$[M_\phi, x] \in \mathcal{K}(\mathbb{H}^d), \quad x \in \Pi.$$

4.1 Principal Symbol on Compact Contact Manifolds

Since $S_i \in \Pi$, it follows that

$$\mathrm{Ext}_{\Omega_i}\left(M_{|J_{\Phi_{i,j}}|^{\frac{1}{2}}} \cdot \mathrm{Rest}_{\Omega_i}(S_i) \cdot M_{|J_{\Phi_{i,j}}|^{-\frac{1}{2}}}\right) \in S_i + \mathcal{K}(L_2(\mathbb{H}^d)).$$

In other words,

$$M_{|J_{\Phi_{i,j}}|^{\frac{1}{2}}} \cdot \mathrm{Rest}_{\Omega_i}(S_i) \cdot M_{|J_{\Phi_{i,j}}|^{-\frac{1}{2}}} \in \mathrm{Rest}_{\Omega_i}(S_i) + \mathcal{K}(L_2(\Omega_i)). \quad (4.1.4)$$

Combining (4.1.3) and (4.1.4), we conclude that

$$\mathrm{Rest}_{\Omega_j}(S_j) \in U_{\Phi_{i,j}}^{-1} \cdot \mathrm{Rest}_{\Omega_i}(S_i) \cdot U_{\Phi_{i,j}} + \mathcal{K}(L_2(\Omega_j)).$$

Thus,

$$S_j \in \mathrm{Ext}_{\Omega_j}\left(U_{\Phi_{i,j}}^{-1} \cdot \mathrm{Rest}_{\Omega_i}(S_i) \cdot U_{\Phi_{i,j}}\right) + \mathcal{K}(L_2(\mathbb{R}^d)).$$

By Theorem 3.3.13, we have

$$\mathrm{Ext}_{\Omega_j}\left(U_{\Phi_{i,j}}^{-1} \cdot \mathrm{Rest}_{\Omega_i}(S_i) \cdot U_{\Phi_{i,j}}\right) \in \Pi$$

and

$$\mathrm{sym}\left(\mathrm{Ext}_{\Omega_j}\left(U_{\Phi_{i,j}}^{-1} \cdot \mathrm{Rest}_{\Omega_i}(S_i) \cdot U_{\Phi_{i,j}}\right)\right) = \pi_{A^{\Phi_{i,j}}}(\mathrm{sym}(S_i)) \circ \Phi_{i,j}^{-1}.$$

By Theorem 1.3.2, compact operators belong to Π. Therefore, $S_j \in \Pi$ and

$$\mathrm{sym}(S_j) \circ h_j = \pi_{A^{\Phi_{i,j}}}(\mathrm{sym}(S_i)) \circ \Phi_{i,j}^{-1} \circ h_j$$
$$= \pi_{A^{\Phi_{i,j}}}(\mathrm{sym}(S_i)) \circ h_i$$
$$= \pi_{A^{\Phi_{i,j}} \circ h_i}(\mathrm{sym}(S_i) \circ h_i)$$
$$= \pi_{i,j}(\mathrm{sym}(S_i) \circ h_i).$$

Finally,

$$\mathrm{sym}_j(S) = \Xi_j(\mathrm{sym}(S_j) \circ h_j)$$
$$= (\Xi_j \circ \pi_{i,j})(\mathrm{sym}(S_i) \circ h_i)$$
$$= \Xi_i(\mathrm{sym}(S_i) \circ h_i)$$
$$= \mathrm{sym}_i(S).$$

$\square$

Proof of Theorem 4.1.8 (i) The condition (4.1.1) in Definition 4.1.1 is verified in Lemma 4.1.10. The condition (ii) in Definition 4.1.1 is immediate. The condition (iii) in Definition 4.1.1 is verified in Lemma 4.1.11.

Let us verify the condition (iv) in Definition 4.1.1. If $i \in \mathbb{I}$ and $S \in \mathcal{K}(L_2(X, \nu))$ is compactly supported in $\mathcal{U}_i$, then $\text{Ext}_{\Omega_i}(W_i S W_i^{-1}) \in \mathcal{K}(L_2(\mathbb{H}^d))$. Using Theorem 1.3.2, we conclude that $\text{Ext}_{\Omega_i}(W_i S W_i^{-1}) \in \Pi$. In other words, $T \in \Pi_i$.

The condition (v) in Definition 4.1.1 is immediate.

Let us verify the condition (vi) in Definition 4.1.1. Let $i \in \mathbb{I}$ and let $\phi \in C_c(\mathcal{U}_i)$. Suppose $\{S_n\}_{n \geq 1} \subset M_\phi \Pi_i M_\phi$ are such that $S_n \to S$ in the uniform norm. It follows that S is compactly supported in $\mathcal{U}_i$ and

$$\text{Ext}_{\Omega_i}(W_i S_n W_i^{-1}) \to \text{Ext}_{\Omega_i}(W_i S W_i^{-1}), \quad n \to \infty,$$

in the uniform norm. The sequence on the left hand side is in Π. Hence, so is its limit. In other words, $S \in \Pi_i$.

Let us verify the condition (vii) in Definition 4.1.1. Let $i \in \mathbb{I}$ and let $S \in \Pi_i$ and $\phi \in C_c(\mathcal{U}_i)$. Let $\psi = \phi \circ h_i^{-1} \in C_c(\mathbb{H}^d)$. We have

$$\text{Ext}_{\Omega_i}(W_i[S, M_\phi]W_i^{-1}) = [\text{Ext}_{\Omega_i}(W_i S W_i^{-1}), M_\psi].$$

Since $\text{Ext}_{\Omega_i}(W_i S W_i^{-1}) \in \Pi$, it follows that the commutator on the right hand side is compact by Lemma 2.2.1. Therefore, the operator on the left hand side is compact and, therefore, so is $[S, M_\phi]$. □

Proof of Theorem 4.1.8 (ii) The condition (i) in Definition 4.1.3 is verified in Lemma 4.1.10. The condition (ii) in Definition 4.1.3 is verified in Lemma 4.1.11.

Let us verify the condition (iii) in Definition 4.1.3. If $S \in \Pi_i$ is compact, then so is $\text{Ext}_{\Omega_i}(W_i S W_i^{-1})$. Since sym vanishes on compact operators, it follows that

$$\text{sym}_i(S) \stackrel{D.4.1.7}{=} \Xi_i \left(\text{sym}(\text{Ext}_{\Omega_i}(W_i S W_i^{-1})) \circ h_i \right) = \Xi_i(0 \circ h_i) = 0.$$

Conversely, if $S \in \Pi_i$ is such that $\text{sym}_i(S) = 0$, then

$$\text{sym}(\text{Ext}_{\Omega_i}(W_i S W_i^{-1})) = 0.$$

Since $\ker(\text{sym}) = \mathcal{K}(L_2(\mathbb{H}^d))$, it follows that

$$\text{Ext}_{\Omega_i}(W_i S W_i^{-1}) \in \mathcal{K}(L_2(\mathbb{H}^d)).$$

Thus, $S \in \mathcal{K}(L_2(X, \nu))$.

The condition (iv) in Definition 4.1.3 is immediate if we take Hom to be the natural embedding $C(X) \to C(E_{\text{hom}})$, which takes any $f \in C(X)$ to a family

4.1 Principal Symbol on Compact Contact Manifolds

$F = \{F_i \in C(\mathcal{U}_i, C^*(\{R_k\}_{k=1}^{2d}))\}_{i\in \mathbb{I}}$ given by $F_i = f\big|_{\mathcal{U}_i} \cdot 1$, where 1 is the unit in $C^*(\{R_k\}_{k=1}^{2d})$. □

Proof of Theorem 1.3.18 By Definition 4.1.9, Π_X is a C^*-algebra and the mapping $\mathrm{sym}_X : \Pi_X \to C(E_{\mathrm{hom}})$ is a $*$-homomorphism. By Definition 4.1.9 and Theorem 4.1.4 (ii), $\ker(\mathrm{sym}_X) = \mathcal{K}(L_2(X, \nu))$.

Let us show that sym_X is surjective. Let $(\phi_n)_{n=1}^N$ be a good[2] partition of unity so that $\phi_n \in C_c^\infty(\mathcal{U}_{i_n})$ for $1 \le n \le N$. Let $F \in C(E_{\mathrm{hom}})$. It follows that

$$q_n = (F\phi_n)_{i_n} \in C_c(\mathcal{U}_{i_n}, C^*(\{R_k\}_{k=1}^{2d})).$$

Since sym is surjective, it follows that there exists $y_n \in \Pi$ such that $\mathrm{sym}(y_n) = q_n \circ h_{i_n}^{-1}$. Let $\psi_n \in C_c(\Omega_{i_n})$ be such that $\phi_n \circ h_{i_n}^{-1} = (\phi_n \circ h_{i_n}^{-1})\psi_n$. We have $x_n = M_{\psi_n} y_n M_{\psi_n} \in \Pi$ and

$$\mathrm{sym}(x_n) = (q_n \circ h_{i_n}^{-1}) \cdot \psi_n^2 = q_n \circ h_{i_n}^{-1}.$$

Since x_n is (bounded and) compactly supported in Ω_{i_n}, it follows that the operator

$$S_n = W_{i_n}^{-1} \mathrm{Rest}_{\Omega_i}(x_n) W_{i_n}$$

is bounded and compactly supported in $\mathcal{U}_{i_n}$. It is clear that $S_n \in \Pi_{i_n} \subset \Pi_X$ and that

$$\mathrm{sym}_X(S_n) = \mathrm{sym}_{i_n}(S_n)$$
$$= \Xi_{i_n}\left(\mathrm{sym}(\mathrm{Ext}_{\Omega_{i_n}}(W_{i_n} S_n W_{i_n}^{-1})) \circ h_{i_n}\right)$$
$$= \Xi_{i_n}(\mathrm{sym}(x_n) \circ h_{i_n})$$
$$= \Xi_{i_n}(q_n)$$
$$= \Xi_{i_n}((F\phi_n)_{i_n})$$
$$= F\phi_n.$$

Thus, $S = \sum_{n=1}^N S_n \in \Pi_X$ and

$$\mathrm{sym}_X(S) = \sum_{n=1}^N \mathrm{sym}_X(S_n) = \sum_{n=1}^N F\phi_n = F.$$

Hence, sym_X is surjective. □

---

[2] See Definition 4.2.1.

4.2 Sub-Laplacian on a Compact Contact Manifold

4.2.1 Sobolev Spaces on Compact Contact Manifolds

If X is a compact manifold, then the linear space $L_2(X, \nu)$ does not depend on the choice of a smooth positive density ν and is, therefore, denoted by $L_2(X)$. The norms on $L_2(X)$ associated to different (smooth positive) densities are equivalent. Similarly, the Sobolev spaces defined below do not depend on the choice of density.

Definition 4.2.1 Let $(\phi_n)_{n=1}^N \subset C^\infty(X)$ be a finite partition of unity. We call it good if each ϕ_n is compactly supported in some chart.

Obviously, good partitions of unity exist only on compact manifolds.

Definition 4.2.2 Let X be a compact contact manifold. Let $(\phi_n)_{n=1}^N$ be a good partition of unity (so that $\phi_n \in C_c^\infty(\mathcal{U}_{i_n})$). We say that a measurable function f on X belongs to the space $W^{m,2}(X)$, $m \in \mathbb{Z}_+$, iff $(f\phi_n) \circ h_{i_n}^{-1} \in W^{m,2}(\mathbb{H}^d)$ for every $1 \leq n \leq N$. We equip $W^{m,2}(X)$ with the norm

$$\|f\|_{W^{m,2}(X)} = \left(\sum_{n=1}^N \|(f\phi_n) \circ h_{i_n}^{-1}\|_{W^{m,2}(\mathbb{H}^d)}^2 \right)^{\frac{1}{2}}, \quad f \in W^{m,2}(X).$$

The space $W^{m,2}(X)$ does not depend on the choice of good partition of unity. The norms associated to different partitions of unity are equivalent.

4.2.2 Definition of Sub-Laplacian

Let G be a sub-Riemannian structure on a compact contact manifold (X, H). For any $i \in \mathbb{I}$, the matrix representation of the metric G in the chart $(\mathcal{U}_i, h_i)$ with respect to the basis $(X_1, \ldots, X_{2d})$ gives rise to a smooth mapping $G_i : \mathcal{U}_i \to \mathrm{GL}^+(2d, \mathbb{R})$. For any $i, j \in \mathbb{I}$ such that $\mathcal{U}_i \cap \mathcal{U}_j \neq \emptyset$, we have

$$G_j(t) = HJ_{\Phi_{j,i}}^*(h_j(t)) \cdot G_i(t) \cdot HJ_{\Phi_{j,i}}(h_j(t)), \quad t \in \mathcal{U}_i \cap \mathcal{U}_j. \tag{4.2.1}$$

Here, $\Phi_{j,i}$ are given in Notation 1.3.4 and $HJ_{\Phi_{j,i}}$ stands for the horizontal Jacobian matrix of $\Phi_{j,i}$ introduced in Definition 1.3.14.

We also denote $g_i = G_i \circ h_i^{-1} : \Omega_i \to \mathrm{GL}^+(2d, \mathbb{R})$, $i \in \mathbb{I}$.

Theorem 4.2.3 *Let (X, H, G) be a compact contact sub-Riemannian manifold with a Heisenberg atlas $\{(\mathcal{U}_i, h_i)\}_{i \in \mathbb{I}}$ and let ν be a smooth positive density on X. For every $i \in \mathbb{I}$, let Ω_i be as in Notation 1.3.4. Let $\Delta_{g_i, \nu \circ h_i^{-1}} : C_c^\infty(\Omega_i) \to C_c^\infty(\Omega_i)$, $i \in \mathbb{I}$, be the sub-Laplacians as in the Definition 1.4.2.*

4.2 Sub-Laplacian on a Compact Contact Manifold

There exists a unique linear operator $\Delta_{G,\nu} : C^\infty(X) \to C^\infty(X)$ *such that*

$$(\Delta_{G,\nu} f) \circ h_i^{-1} = \Delta_{g_i, \nu \circ h_i^{-1}}(f \circ h_i^{-1}), \quad f \in C_c^\infty(\mathcal{U}_i).$$

Theorem 4.2.3 is proved in Sect. 4.2.3. Moreover, we show that $\Delta_{G,\nu}$ does not depend on the choice of Heisenberg atlas.

4.2.3 Proof of Theorem 4.2.3

Lemma 4.2.4 *Let Ω be equipped with a smooth positive density ν and let Ω' be equipped with a smooth positive density ν'. Let $\Phi : \Omega \to \Omega'$ be a Heisenberg diffeomorphism. Suppose $\nu' = \nu \circ \Phi^{-1}$. Suppose $g : \Omega \to \mathrm{GL}^+(2d, \mathbb{R})$ and $g' : \Omega' \to \mathrm{GL}^+(2d, \mathbb{R})$ be smooth mappings such that*

$$g' = HJ_{\Phi^{-1}}^* \cdot (g \circ \Phi^{-1}) \cdot HJ_{\Phi^{-1}}.$$

We have

$$\Delta_{g',\nu'} = V_\Phi^{-1} \Delta_{g,\nu} V_\Phi.$$

Proof We clearly have

$$V_\Phi^{-1} \Delta_{g,\nu} V_\Phi = \sum_{k,l=1}^{2d} V_\Phi^{-1} X_k^* V_\Phi V_\Phi^{-1} M_{(g^{-1})_{k,l}} V_\Phi V_\Phi^{-1} X_l V_\Phi.$$

By Theorem 3.3.1,

$$V_\Phi^{-1} X_l V_\Phi = \sum_{j=1}^{2d} M_{(HJ_\Phi)_{jl} \circ \Phi^{-1}} X_j.$$

The mapping $\Phi : (\Omega, \nu) \to (\Omega', \nu')$ is bijective and preserves the measure. Hence, the operator $V_\Phi : L_2(\Omega', \nu') \to L_2(\Omega, \nu)$ is unitary. Taking the adjoints of both parts in the last equality, we get

$$V_\Phi^{-1} X_k^* V_\Phi = \sum_{i=1}^{2d} X_i^* M_{(HJ_\Phi)_{ik} \circ \Phi^{-1}}.$$

Here X_k^* on the left hand side is the adjoint in $L_2(\Omega', \nu')$, whereas X_i^* on the right hand side is the adjoint in $L_2(\Omega, \nu)$. Clearly, $V_\Phi^{-1} M_f V_\Phi = M_{f \circ \Phi^{-1}}$ for every function f. We conclude that

$$V_\Phi^{-1} \Delta_g V_\Phi = \sum_{i,j,k,l=1}^{2d} X_i^* M_{(HJ_\Phi)_{ik} \circ \Phi^{-1}} M_{(g^{-1})_{k,l} \circ \Phi^{-1}} M_{(HJ_\Phi)_{jl} \circ \Phi^{-1}} X_j.$$

Thus, we need to establish the equalities

$$M_{(g^{-1})_{i,j}} = \sum_{k,l=1}^{2d} M_{(HJ_\Phi)_{ik} \circ \Phi^{-1}} M_{(g^{-1})_{k,l} \circ \Phi^{-1}} M_{(HJ_\Phi)_{jl} \circ \Phi^{-1}}, \quad 1 \le i, j \le 2d.$$

In other words, we need to verify

$$(g')^{-1} = \left(HJ_\Phi \cdot g^{-1} \cdot (HJ_\Phi)^* \right) \circ \Phi^{-1}$$

pointwise on Ω'. Taking the inverses, we rewrite the equality as

$$g' = \left((HJ_\Phi)^* \circ \Phi^{-1} \right)^{-1} \cdot (g \circ \Phi^{-1}) \cdot \left(HJ_\Phi \circ \Phi^{-1} \right)^{-1}.$$

It follows from the composition rule (1.3.4) that

$$\left(HJ_\Phi \circ \Phi^{-1} \right)^{-1} = HJ_{\Phi^{-1}}.$$

The assertion follows now from the relation between g and g'. □

Definition 4.2.5 Let (X, H, G) be a compact contact sub-Riemannian manifold and let ν be a smooth positive density on X. For every $i \in \mathbb{I}$, let Ω_i be as in Notation 1.3.4.

Let $\Delta_{g_i} : C_c^\infty(\Omega_i) \to C_c^\infty(\Omega_i)$, $i \in \mathbb{I}$, be the sub-Laplacian as in the Definition 1.4.2. Let $(\phi_n)_{n=1}^N$ be a good partition of unity (so that $\phi_n \in C_c^\infty(\mathcal{U}_{i_n})$).

The sub-Laplacian $\Delta_{G,\nu} : C^\infty(X) \to C^\infty(X)$ is defined by the formula

$$\Delta_{G,\nu} f = \sum_{n=1}^N \left(\Delta_{g_{i_n}, \nu \circ h_{i_n}^{-1}} ((f \phi_n) \circ h_{i_n}^{-1}) \right) \circ h_{i_n}, \quad f \in C^\infty(X).$$

The next lemma completes the proof of Theorem 4.2.3.

Lemma 4.2.6 *The operator $\Delta_{G,\nu}$ does not depend on the particular choice of a good partition of unity and satisfies the condition*

$$\Delta_{G,\nu} f = \left(\Delta_{g_i, \nu \circ h_i^{-1}} (f \circ h_i^{-1}) \right) \circ h_i, \quad f \in C_c^\infty(\mathcal{U}_i). \tag{4.2.2}$$

4.2 Sub-Laplacian on a Compact Contact Manifold

Proof Let $\mathcal{U}_i \cap \mathcal{U}_j \neq \varnothing$. If $f \in C_c^\infty(\mathcal{U}_i \cap \mathcal{U}_j)$, then we claim that

$$\left(\Delta_{g_i, \nu \circ h_i^{-1}}(f \circ h_i^{-1})\right) \circ h_i = \left(\Delta_{g_j, \nu \circ h_j^{-1}}(f \circ h_j^{-1})\right) \circ h_j. \tag{4.2.3}$$

It is easy to see that (4.2.3) is equivalent to

$$\Delta_{g_j, \nu \circ h_j^{-1}} = V_{\Phi_{i,j}}^{-1} \Delta_{g_i, \nu \circ h_i^{-1}} V_{\Phi_{i,j}}. \tag{4.2.4}$$

To prove (4.2.4), we apply Lemma 4.2.4 to the measure spaces $(\Omega_{i,j}, \nu \circ h_i^{-1})$, $(\Omega_{j,i}, \nu \circ h_j^{-1})$ and the diffeomorphism $\Phi_{i,j}$. Let us check that they satisfy the equalities (on $\Omega_{j,i}$)

$$(\nu \circ h_j^{-1}) = (\nu \circ h_i^{-1}) \circ \Phi_{i,j}^{-1}, \quad g_j = HJ^*_{\Phi_{i,j}^{-1}} \cdot (g_i \circ \Phi_{i,j}^{-1}) \cdot HJ_{\Phi_{i,j}^{-1}} \tag{4.2.5}$$

required in Lemma 4.2.4. The first equality in (4.2.5) is obvious. Note that (on $\Omega_{j,i}$)

$$g_j = G_j \circ h_j^{-1}, \quad g_i \circ \Phi_{i,j}^{-1} = G_i \circ h_i^{-1} \circ (h_i \circ h_j^{-1}) = G_i \circ h_j^{-1}.$$

Composing with h_j, we derive the second equality in (4.2.5) from (4.2.1). By Lemma 4.2.4, we get (4.2.4) and, therefore, (4.2.3).

Now, fix $i \in \mathbb{I}$ and $f \in C_c^\infty(\mathcal{U}_i)$. Then $f\phi_n \in C_c^\infty(\mathcal{U}_i \cap \mathcal{U}_{i_n})$. By (4.2.3), we get

$$\left(\Delta_{g_{i_n}, \nu \circ h_{i_n}^{-1}}((f\phi_n) \circ h_{i_n}^{-1})\right) \circ h_{i_n} = \left(\Delta_{g_i, \nu \circ h_i^{-1}}((f\phi_n) \circ h_i^{-1})\right) \circ h_i.$$

Summing over $1 \leq n \leq N$, we infer (4.2.2).

Let now $(\psi_m)_{m=1}^M$ be another partition of unity. Suppose $\psi_m \in C_c^\infty(\mathcal{U}_{j_m})$. Using (4.2.2) with $i = j_m$, we obtain

$$\left(\Delta_{g_{j_m}, \nu \circ h_{j_m}^{-1}}((f\psi_m) \circ h_{j_m}^{-1})\right) \circ h_{j_m} = \Delta_{G,\nu}(f\psi_m), \quad f \in C^\infty(X).$$

Summing over $1 \leq m \leq M$, we obtain

$$\sum_{m=1}^M \left(\Delta_{g_{j_m}, \nu \circ h_{j_m}^{-1}}((f\psi_m) \circ h_{j_m}^{-1})\right) \circ h_{j_m} = \Delta_{G,\nu}(f), \quad f \in C^\infty(X).$$

This demonstrates that the operator $\Delta_{G,\nu}$ does not depend on the choice of good partition of unity. □

Recall that two Heisenberg atlases $(\mathcal{U}_i, h_i)_{i \in \mathbb{I}}$ and $(\mathcal{U}'_j, h'_j)_{j \in \mathbb{J}}$ are equivalent if their union $(\mathcal{U}_i, h_i)_{i \in \mathbb{I}} \cup (\mathcal{U}'_j, h'_j)_{j \in \mathbb{J}}$ is a Heisenberg atlas. Assume that the operators $\Delta_{G,\nu}$ and $\Delta'_{G,\nu}$ are defined by equivalent Heisenberg atlases $(\mathcal{U}_i, h_i)_{i \in \mathbb{I}}$

and $(\mathcal{U}'_j, h'_j)_{j \in \mathbb{J}}$, respectively. Applying Lemma 4.2.6 to $(\mathcal{U}_i, h_i)_{i \in \mathbb{I}} \cup (\mathcal{U}'_j, h'_j)_{j \in \mathbb{J}}$, we immediately infer that $\Delta_{G,\nu} = \Delta'_{G,\nu}$. So the operator $\Delta_{G,\nu}$ does not depend on the particular choice of a Heisenberg atlas.

4.2.4 Ellipticity of the Sub-Laplacian

In this subsection, we prove an analogue of the standard elliptic estimate for the sub-Laplacian $\Delta_{G,\nu}$.

Theorem 4.2.7 *Let (X, H, G) be a compact contact sub-Riemannian manifold and let ν be a smooth positive density on X. We have*

$$\|\Delta_{G,\nu} f\|_{L_2(X,\nu)} + \|f\|_{L_2(X,\nu)} \geq c_{X,G,\nu} \|f\|_{W^{2,2}(X)}, \quad f \in W^{2,2}(X).$$

The rest of the subsection is devoted to the proof of this theorem.

Lemma 4.2.8 *Let (X, H, G) be a compact contact sub-Riemannian manifold and let ν be a smooth positive density on X. Let $(\mathcal{U}, h)$ be a chart and let $K \subset \mathcal{U}$ be a compact set. If $f \in W^{2,2}(X)$ is supported in K, then*

$$\|\Delta_{G,\nu} f\|_{L_2(X,\nu)} + \|f\|_{L_2(X,\nu)} \geq c_{\mathcal{U},h,K,G,\nu} \|f\|_{W^{2,2}(X)}.$$

Proof Choose a differential operator P of order 2 on $\mathbb{H}^d$ such that

(1) if $F \in W^{2,2}(X)$ is supported on K, then

$$\Delta_{G,\nu} F = \left(P\left(F \circ h^{-1} \right) \right) \circ h;$$

(2) the coefficients of P are smooth and constant outside some neighborhood of K;
(3) P is formally self-adjoint and formally positive on $W^{2,2}(\mathbb{H}^d)$;
(4) P is formally elliptic.

Applying Lemma 2.6.4 with $m = 1$, we obtain that P is uniformly elliptic. In particular, P satisfies the assumptions in Theorem 2.4.3.

Since f is compactly supported in $\mathcal{U}$, it follows that

$$\|\Delta_{G,\nu} f\|_{L_2(X,\nu)} = \|P(f \circ h^{-1})\|_{L_2(h(\mathcal{U}), \nu \circ h^{-1})} \geq c^{(1)}_{\mathcal{U},h,K,\nu} \|P(f \circ h^{-1})\|_{L_2(\mathbb{H}^d)}.$$

Also,

$$\|f\|_{L_2(X,\nu)} \geq c^{(1)}_{\mathcal{U},h,K,\nu} \|f \circ h^{-1}\|_{L_2(\mathbb{H}^d)}.$$

4.2 Sub-Laplacian on a Compact Contact Manifold

Thus,

$$\|\Delta_{G,\nu} f\|_{L_2(X,\nu)} + \|f\|_{L_2(X,\nu)} \geq c^{(1)}_{\mathcal{U},h,K,\nu}\left(\|P(f\circ h^{-1})\|_{L_2(\mathbb{H}^d)} + \|f\circ h^{-1}\|_{L_2(\mathbb{H}^d)}\right).$$

By Theorem 2.4.3, we have

$$\|P(f\circ h^{-1})\|_{L_2(\mathbb{H}^d)} + \|f\circ h^{-1}\|_{L_2(\mathbb{H}^d)} \geq c_P \|f\circ h^{-1}\|_{W^{2,2}(\mathbb{H}^d)}.$$

Note that

$$\|f\circ h^{-1}\|_{W^{2,2}(\mathbb{H}^d)} \geq c^{(2)}_{\mathcal{U},h,K,\nu}\|f\|_{W^{2,2}(X)}.$$

Combining the last 3 inequalities, we write

$$\|\Delta_{G,\nu} f\|_{L_2(X,\nu)} + \|f\|_{L_2(X,\nu)} \geq c^{(1)}_{\mathcal{U},h,K,\nu} c^{(2)}_{\mathcal{U},h,K,\nu} c_P \|f\|_{W^{2,2}(X)}.$$

This completes the proof. □

Lemma 4.2.9 *We have*

$$\|f\|_{W^{1,2}(X)} \leq c_{\text{abs}} \|f\|^{\frac{1}{2}}_{L_2(X)} \|f\|^{\frac{1}{2}}_{W^{2,2}(X)}, \quad f \in W^{2,2}(X).$$

Proof Let $(\phi_n)_{n=1}^N$ be a good partition of unity as in Definition 4.2.2. By complex interpolation (for the Sobolev spaces on $\mathbb{H}^d$), we have

$$\|(f\phi_n)\circ h^{-1}_{i_n}\|_{W^{1,2}(\mathbb{H}^d)} \leq c_{\text{abs}} \|(f\phi_n)\circ h^{-1}_{i_n}\|^{\frac{1}{2}}_{L_2(\mathbb{H}^d)} \|(f\phi_n)\circ h^{-1}_{i_n}\|^{\frac{1}{2}}_{W^{2,2}(\mathbb{H}^d)}.$$

Thus,

$$\|f\|^2_{W^{1,2}(X)} = \sum_{n=1}^N \|(f\phi_n)\circ h^{-1}_{i_n}\|^2_{W^{1,2}(\mathbb{H}^d)}$$

$$\leq c_{\text{abs}} \sum_{n=1}^N \|(f\phi_n)\circ h^{-1}_{i_n}\|_{L_2(\mathbb{H}^d)} \|(f\phi_n)\circ h^{-1}_{i_n}\|_{W^{2,2}(\mathbb{H}^d)}$$

$$\leq c_{\text{abs}} \left(\sum_{n=1}^N \|(f\phi_n)\circ h^{-1}_{i_n}\|^2_{L_2(\mathbb{H}^d)}\right)^{\frac{1}{2}} \left(\sum_{n=1}^N \|(f\phi_n)\circ h^{-1}_{i_n}\|^2_{W^{2,2}(\mathbb{H}^d)}\right)^{\frac{1}{2}}$$

$$= c_{\text{abs}} \|f\|^{\frac{1}{2}}_{L_2(X)} \|f\|^{\frac{1}{2}}_{W^{2,2}(X)}.$$

□

Proof of Theorem 4.2.7 Let $(\phi_n)_{n=1}^N$ be a good partition of unity (so that $\phi_n \in C_c^\infty(\mathcal{U}_{i_n})$). Using triangle inequality, we write

$$\|\Delta_{G,\nu} f\|_{L_2(X,\nu)} \geq \|M_{\phi_n} \Delta_{G,\nu} f\|_{L_2(X,\nu)}$$
$$\geq \|\Delta_{G,\nu}(f\phi_n)\|_{L_2(X,\nu)} - \|[M_{\phi_n}, \Delta_{G,\nu}]f\|_{L_2(X,\nu)}.$$

By Lemma 4.2.8, we have

$$\|\Delta_{G,\nu}(f\phi_n)\|_{L_2(X,\nu)} + \|f\phi_n\|_{L_2(X,\nu)} \geq c_{\mathcal{U}_{i_n}, h_{i_n}, \mathrm{supp}(\phi_n), G, \nu} \|f\phi_n\|_{W^{2,2}(X)}.$$

Since $[M_{\phi_n}, \Delta_{G,\nu}]$ is a differential operator of order 1, it follows that

$$\|[M_{\phi_n}, \Delta_{G,\nu}]f\|_{L_2(X,\nu)} \leq \|[M_{\phi_n}, \Delta_{G,\nu}]\|_{W^{1,2}(X) \to L_2(X)} \|f\|_{W^{1,2}(X)}.$$

Setting

$$c_{X,G,\nu}^{(1)} = \min_{1 \leq n \leq N} c_{\mathcal{U}_{i_n}, h_{i_n}, \mathrm{supp}(\phi_n), G, \nu},$$

$$c_{X,G,\nu}^{(2)} = \max_{1 \leq n \leq N} c_{\mathcal{U}_{i_n}, h_{i_n}, \mathrm{supp}(\phi_n), G, \nu},$$

$$c_{X,G,\nu}^{(3)} = \max_{1 \leq n \leq N} \|[M_{\phi_n}, \Delta_{G,\nu}]\|_{W^{1,2}(X) \to L_2(X)},$$

we write

$$\|\Delta_{G,\nu} f\|_{L_2(X,\nu)} \geq c_{X,G,\nu}^{(1)} \|f\phi_n\|_{W^{2,2}(X)} - c_{X,G,\nu}^{(2)} \|f\phi_n\|_{L_2(X)} - c_{X,G,\nu}^{(3)} \|f\|_{W^{1,2}(X)}.$$

Thus,

$$\|\Delta_{G,\nu} f\|_{L_2(X,\nu)} + (c_{X,G,\nu}^{(2)} + c_{X,G,\nu}^{(3)}) \|f\|_{W^{1,2}(X)} \geq c_{X,G}^{(1)} \|f\phi_n\|_{W^{2,2}(X)}.$$

Taking squares and summing over $1 \leq n \leq N$, we obtain

$$\left(\|\Delta_{G,\nu} f\|_{L_2(X,\nu)} + (c_{X,G,\nu}^{(2)} + c_{X,G,\nu}^{(3)}) \|f\|_{W^{1,2}(X)}\right)^2$$
$$\geq \frac{1}{N}(c_{X,G,\nu}^{(1)})^2 \sum_{n=1}^N \|f\phi_n\|_{W^{2,2}(X)}^2$$
$$\geq \left(c_{X,G,\nu}^{(4)} \|f\|_{W^{2,2}(X)}\right)^2.$$

4.2 Sub-Laplacian on a Compact Contact Manifold 133

Thus,

$$\|\Delta_{G,\nu} f\|_{L_2(X,\nu)} + (c^{(2)}_{X,G,\nu} + c^{(3)}_{X,G,\nu})\|f\|_{W^{1,2}(X)} \geq c^{(4)}_{X,G,\nu}\|f\|_{W^{2,2}(X)}.$$

By Lemma 4.2.9, we have

$$\|\Delta_{G,\nu} f\|_{L_2(X,\nu)} + c^{(5)}_{X,G}\|f\|^{\frac{1}{2}}_{W^{2,2}(X)}\|f\|^{\frac{1}{2}}_{L_2(X)} \geq c^{(4)}_{X,G}\|f\|_{W^{2,2}(X)}.$$

This immediately yields the assertion. □

4.2.5 Self-Adjointness of the Sub-Laplacian

In this subsection, we prove self-adjointness of the sub-Laplacian $\Delta_{G,\nu}$ stated in Theorem 1.4.3.

Proof of Theorem 1.4.3 By Theorem 4.2.7, the operator $\Delta_{G,\nu}$ considered as an unbounded operator in $L_2(X, \nu)$ with domain $W^{2,2}(X, \nu)$ is closed. To show its self-adjointness, it suffices (see Theorem VIII.3 in [57]) to establish that every $u \in \text{dom}(\Delta^*_{G,\nu})$ with $(\Delta^*_{G,\nu} + \iota)u = 0$ (or with $(\Delta^*_{G,\nu} - \iota)u = 0$) is trivial.

Denote by $\mathscr{P}$ the same differential operator $\Delta_{G,\nu}$ acting on distributions. We have $\Delta^*_{G,\nu} = \mathscr{P}|_{\text{dom}(\Delta^*_{G,\nu})}$. If $u \in \text{dom}(\Delta^*_{G,\nu})$ with $(\Delta^*_{G,\nu} + \iota)u = 0$, then $u \in L_2(X, \nu)$ and $(\mathscr{P} + \iota)u = 0$. In other words, u is a weak solution of the equation $(\Delta_{G,\nu} + \iota)u = 0$.

Fix $i \in \mathbb{I}$ and define a distribution u_i on Ω_i by setting

$$\langle u_i, \phi \rangle = \langle u, \phi \circ h_i \rangle, \quad \phi \in C^\infty_c(\Omega_i).$$

By the definition of the action of $\Delta_{G,\nu}$ on distributions, we have

$$\langle u, (\Delta_{G,\nu} + \iota)(\phi \circ h_i) \rangle = \langle (\Delta_{G,\nu} + \iota)u, \phi \circ h_i \rangle = 0.$$

On the other hand, by (4.2.2),

$$(\Delta_{G,\nu} + \iota)(\phi \circ h_i) = \left((\Delta_{g_i, \nu \circ h_i^{-1}} + \iota)\phi\right) \circ h_i.$$

Thus,

$$\langle (\Delta_{g_i, \nu \circ h_i^{-1}} + \iota)u_i, \phi \rangle = \langle u_i, (\Delta_{g_i, \nu \circ h_i^{-1}} + \iota)\phi \rangle = 0, \quad \phi \in C^\infty_c(\Omega_i).$$

In other words, u_i is a weak solution of the equation

$$(\Delta_{g_i, \nu \circ h_i^{-1}} + \iota)u_i = 0.$$

By Lemma 2.5.3, u_i is smooth. Since $i \in \mathbb{I}$ is arbitrary, we infer that u is smooth. Since $\Delta_{G,v}$ is formally self-adjoint, it follows that

$$\langle \Delta_{G,v}u, u \rangle_{L_2(X,v)} = \langle u, \Delta_{G,v}u \rangle_{L_2(X,v)}.$$

Since $\Delta_{G,v}u = -\iota u$, it follows that

$$-\iota \|u\|^2_{L_2(X,v)} = \langle -\iota u, u \rangle_{L_2(X,v)} = \langle u, -\iota u \rangle_{L_2(X,v)} = \iota \|u\|^2_{L_2(X,v)}.$$

Thus, $\|u\|_{L_2(X,v)} = 0$ and $u = 0$. □

4.3 Liouville Trace on the Group von Neumann Algebra Bundle

In this subsection, we prove the existence of the analogue of the Liouville measure in the contact setting stated in Theorem 1.4.4. Recall that τ denotes the natural trace on $\text{VN}(\mathbb{H}^d)$ (see Sect. 2.1.3).

Lemma 4.3.1 *If $A \in \text{Aut}(\mathbb{H}^d)$, then*

$$\tau(\pi_A(x)) = \det(A)^{-1} \cdot \tau(x), \quad x \in L_1(\text{VN}(\mathbb{H}^d)).$$

Proof It is sufficient to consider the case when $x = \lambda(f)$ with $f \in C_c(\mathbb{H}^d)$. Then, for $\xi \in L_2(\mathbb{H}^d)$ and $h \in \mathbb{H}^d$, we have

$$(\pi_A(x)\xi)(h) = [V_A^{-1}\lambda(f)V_A]\xi(h)$$
$$= (\lambda(f)[V_A\xi])((A^*)^{-1}h)$$
$$= \int_{\mathbb{H}^d} f((A^*)^{-1}(h)g^{-1})\xi(A^*(g))dg.$$

Now we make the linear change of variables $g_1 = A^*g$ in the integral:

$$\pi_A(x)\xi(h) = \det(A)^{-1} \int_{\mathbb{H}^d} f((A^*)^{-1}h \cdot ((A^*)^{-1}g_1)^{-1})\xi(g_1)dg_1.$$

Since $(A^*)^{-1} \in \text{Aut}(\mathbb{H}^d)$, it follows that

$$(A^*)^{-1}h \cdot ((A^*)^{-1}g_1)^{-1} = (A^*)^{-1}(hg_1^{-1}).$$

4.3 Liouville Trace on the Group von Neumann Algebra Bundle

Thus,

$$(\pi_A(x)\xi)(h) = \det(A)^{-1} \int_{\mathbb{H}^d} (f \circ (A^*)^{-1})(hg_1^{-1})\xi(g_1)dg_1$$
$$= \det(A)^{-1}[\lambda(V_{A^{-1}}f)\xi](h).$$

We infer that

$$\tau(\pi_A(x)) = \det(A)^{-1}(V_{A^{-1}}f)(0) = \det(A)^{-1}f(0) = \det(A)^{-1}\tau(x),$$

as desired. □

Lemma 4.3.2 *Let $\Omega, \Omega' \subset \mathbb{H}^d$ be connected open sets. Let $\Phi : \Omega \to \Omega'$ be a Heisenberg diffeomorphism. For every $x \in L_\infty(\Omega, \mathrm{VN}(\mathbb{H}^d))$, we have*

$$(\int_\Omega \otimes \tau)(x) = (\int_{\Omega'} \otimes \tau)(\pi_{A^\Phi}(x) \circ \Phi^{-1}).$$

Proof If we write $x : p \mapsto x(p)$, $p \in \Omega$, then the desired equation can be written as

$$\int_\Omega \tau(x(p))dp = \int_{\Omega'} \tau(\pi_{A^\Phi(\Phi^{-1}(p'))}(x(\Phi^{-1}(p'))))dp'.$$

Substituting $p' = \Phi(p)$ in the integral on the right hand side, we rewrite the equality as follows:

$$\int_\Omega \tau(x(p))dp = \int_\Omega \tau(\pi_{A^\Phi(p)}(x(p)))J_\Phi(p)dp.$$

By Lemma 4.3.1, we have

$$\tau(\pi_{A^\Phi(p)}(x(p))) = \det(A^\Phi(p))^{-1} \cdot \tau(x(p)).$$

Using that $\det J_\Phi(p) = (\lambda_\Phi(p))^{d+1}$ (see e.g. [42, p.34]) and Theorem 2.1.6, we get the equality $\det(A^\Phi(p)) = \det(J_\Phi(p))$, which completes the proof. □

Proof of Theorem 1.4.4 Fix a sequence $\{A_n\}_{n\geq 1}$ of pairwise disjoint Borel measurable sets such that $\cup_{n\geq 1} A_n = X$ and $A_n \subset \mathcal{U}_{i_n}$ for every $n \geq 1$.

There is a natural von Neumann algebra $*$-isomorphism

$$\pi_{\mathrm{coord}} : L_\infty(E) \to \bigoplus_{n\geq 1} L_\infty(h_{i_n}(A_n), \mathrm{VN}(\mathbb{H}^d))$$

given by the formula

$$\pi_{\mathrm{coord}} : F \mapsto \bigoplus_{n\geq 1}(F_{i_n}\chi_{A_n}) \circ h_{i_n}^{-1}.$$

Define a faithful normal semifinite trace τ_n on $L_\infty(h_{i_n}(A_n), \text{VN}(\mathbb{H}^d))$ by setting

$$\tau_n(x) = \int_{h_{i_n}(A_n)} \tau(x(p)) dp, \quad x \in L_\infty(h_{i_n}(A_n), \text{VN}(\mathbb{H}^d)).$$

We now define a linear functional Λ on $L_\infty(E)$ by

$$\Lambda(F) = \left(\bigoplus_{n \geq 1} \tau_n\right)(\pi_{\text{coord}}(F)), \quad F \in L_\infty(E).$$

It is immediate that the so-defined Λ is a faithful normal semifinite weight on $L_\infty(E)$. Since χ_{A_n} is a central element of $L_\infty(E)$ for every $n \geq 1$, the weight Λ is unitarily invariant and is, therefore, a faithful normal semifinite trace on $L_\infty(E)$. The equality (1.4.1) follows now from Lemma 4.3.2. □

4.4 Proof of Theorem 1.4.6

In this section, we prove Theorem 1.4.6. First, we prove the existence of the canonical weight stated in Lemma 1.4.5.

Let (X, H, G) be a compact contact sub-Riemannian manifold and let ν be a smooth positive density on X. Recall that we consider a locally trivial bundle of Hilbert spaces $\mathcal{H} = (X, L_2(\mathbb{H}^d), \nu)$ on X with fiber $L_2(\mathbb{H}^d)$ defined by the family $\upsilon = \{\upsilon_{i,j}\}_{i,j \in \mathbb{I}}$ of continuous mappings $\upsilon_{i,j} : \mathcal{U}_i \cap \mathcal{U}_j \to U(L_2(\mathbb{H}^d))$ (here, $U(L_2(\mathbb{H}^d))$ is the group of all unitary operators on the Hilbert space $L_2(\mathbb{H}^d)$) given on each nonempty overlap $\mathcal{U}_i \cap \mathcal{U}_j$ by $\upsilon_{i,j} = V_{A^{\Phi_{i,j}} \circ h_i}$.

Definition 4.4.1 The Hilbert space $L_2(X, \mathcal{H}, \nu)$ of L_2-sections of $\mathcal{H}$ is the set of all families $\xi = \{\xi_i\}_{i \in \mathbb{I}}$ such that

(i) for every $i \in \mathbb{I}$, we have $\xi_i \in L_2(\mathcal{U}_i, L_2(\mathbb{H}^d))$;
(ii) for every $i, j \in \mathbb{I}$, we have $\xi_i = \upsilon_{i,j}(\xi_j)$ on $\mathcal{U}_i \cap \mathcal{U}_j$;
(iii)

$$\|\xi\|^2_{L_2(X, \mathcal{H}, \nu)} := \int_X \|\xi(x)\|^2_{L_2(\mathbb{H}^d)} d\nu(x) < \infty.$$

Here for any $x \in X$, the norm $\|\xi(x)\|_{L_2(\mathbb{H}^d)}$ is defined by

$$\|\xi(x)\|_{L_2(\mathbb{H}^d)} := \|\xi_i(x)\|_{L_2(\mathbb{H}^d)}, \quad x \in \mathcal{U}_i.$$

It it clear that the right-hand side is independent of the choice of i.

4.4 Proof of Theorem 1.4.6

We consider the C^*-algebra bundle $E = (X, \text{VN}(\mathbb{H}^d), \varpi)$ associated with the family $\varpi = \{\pi_{i,j}\}_{i,j \in \mathbb{I}}$ of continuous mappings $\pi_{i,j} : \mathcal{U}_i \cap \mathcal{U}_j \to \text{Aut}(\text{VN}(\mathbb{H}^d))$ defined on each nonempty overlap $\mathcal{U}_i \cap \mathcal{U}_j$ by $\pi_{i,j} = \pi_{A^{\Phi_{i,j} \circ h_i}}$. There is a natural faithful $*$-representation of the von Neumann algebra $L_\infty(E)$ of its bounded measurable sections on $L_2(X, \mathcal{H}, \nu)$. For $F = \{F_i \in L_\infty(\mathcal{U}_i, \text{VN}(\mathbb{H}^d))\}_{i \in \mathbb{I}} \in L_\infty(E)$ and $\xi = \{\xi_i \in L_2(\mathcal{U}_i, L_2(\mathbb{H}^d))\}_{i \in \mathbb{I}}$, we have $F\xi = \{F_i\xi_i \in L_2(\mathcal{U}_i, L_2(\mathbb{H}^d))\}_{i \in \mathbb{I}}$, where $F_i\xi_i$ is given by the pointwise action of $\text{VN}(\mathbb{H}^d)$ in $L_2(\mathbb{H}^d)$.

Proof of Lemma 1.4.5 For any $t \in \mathcal{U}_i$, the operator

$$-\sum_{k_1,k_2=1}^{2d} (G_i^{-1}(t))_{k_1,k_2} X_{k_1} X_{k_2}$$

is an unbounded self-adjoint positive operator on $L_2(\mathbb{H}^d)$ affiliated with $\text{VN}(\mathbb{H}^d)$. The family of such operators gives rise to an unbounded self-adjoint positive operator

$$\Delta_i = -\sum_{k_1,k_2=1}^{2d} (G_i^{-1})_{k_1,k_2} \otimes X_{k_1} X_{k_2},$$

on $L_2(\mathcal{U}_i, \mathcal{H}, \nu) = L_2(\mathcal{U}_i, \nu) \otimes L_2(\mathbb{H}^d)$ affiliated with $L_\infty(\mathcal{U}_i, \nu)\bar{\otimes}\text{VN}(\mathbb{H}^d)$. To prove the existence of an unbounded operator q_X on $L_2(X, \mathcal{H}, \nu)$ affiliated with $L_\infty(E)$ such that for every $i \in \mathbb{I}$, its restriction $(q_X)_i$ to $L_2(\mathcal{U}_i, \mathcal{H}, \nu)$ coincides with Δ_i, we need to check that the operators Δ_i satisfy the compatibility condition

$$\Delta_j = \pi_{A^{\Phi_{i,j} \circ h_i}}(\Delta_i) = \pi_{i,j}(\Delta_i). \tag{4.4.1}$$

Once we prove the condition (4.4.1), the definition of q_X is straightforward, and the facts that it is an unbounded self-adjoint positive operator affiliated with $L_\infty(E)$ follow exactly as in the case of direct integrals of Hilbert spaces (see, for instance, [56, Section XIII.16]).

Clearly,

$$\pi_{A^{\Phi_{i,j}}}\left(\sum_{k_1,k_2=1}^{2d} (g_i^{-1})_{k_1,k_2} \otimes X_{k_1} X_{k_2}\right)$$

$$= \sum_{k_1,k_2=1}^{2d} ((g_i^{-1})_{k_1,k_2} \otimes 1) \cdot \pi_{A^{\Phi_{i,j}}}(1 \otimes X_{k_1}) \cdot \pi_{A_{i,j}^{\Phi}}(1 \otimes X_{k_2}).$$

By (2.1.7), we have

$$\pi_{A^{\Phi_{i,j}}}(1 \otimes X_k) = \sum_{l=1}^{2d}(A^{\Phi_{i,j}})_{kl} \otimes X_l.$$

Thus,

$$\pi_{A^{\Phi_{i,j}}}\left(\sum_{k_1,k_2=1}^{2d}(g_i^{-1})_{k_1,k_2} \otimes X_{k_1}X_{k_2}\right)$$

$$= \sum_{k_1,k_2=1}^{2d}\sum_{l_1,l_2=1}^{2d}(A^{\Phi_{i,j}})_{k_1 l_1}((g_i^{-1})_{k_1,k_2}(A^{\Phi_{i,j}})_{k_2 l_2} \otimes X_{l_1}X_{l_2}$$

$$= \sum_{l_1,l_2=1}^{2d}\left(\sum_{k_1,k_2=1}^{2d}(A^{\Phi_{i,j}})_{k_1 l_1}(g_i^{-1})_{k_1,k_2}(A^{\Phi_{i,j}})_{k_2 l_2}\right) \otimes X_{l_1}X_{l_2}.$$

It is immediate that

$$\sum_{k_1,k_2=1}^{2d}(A^{\Phi_{i,j}})_{k_1 l_1}(g_i^{-1})_{k_1,k_2}(A^{\Phi_{i,j}})_{k_2 l_2} = (HJ^*_{\Phi_{i,j}} \cdot g_i^{-1} \cdot HJ_{\Phi_{i,j}})_{l_1 l_2}.$$

Thus,

$$\pi_{A^{\Phi_{i,j}}}\left(\sum_{k_1,k_2=1}^{2d}(g_i^{-1})_{k_1,k_2} \otimes X_{k_1}X_{k_2}\right) = \sum_{l_1,l_2=1}^{2d}(HJ^*_{\Phi_{i,j}} \cdot g_i^{-1} \cdot HJ_{\Phi_{i,j}})_{l_1 l_2} \otimes X_{l_1}X_{l_2}.$$

It follows now from the equality

$$g_j^{-1} = \left(HJ^*_{\Phi_{i,j}} \cdot g_i^{-1} \cdot HJ_{\Phi_{i,j}}\right) \circ \Phi_{i,j}^{-1}$$

that

$$\sum_{k_1,k_2=1}^{2d}(g_j^{-1})_{k_1,k_2} \otimes X_{k_1}X_{k_2} = \left(\pi_{A^{\Phi}_{i,j}}\left(\sum_{k_1,k_2=1}^{2d}(g_i^{-1})_{k_1,k_2} \otimes X_{k_1}X_{k_2}\right)\right) \circ \Phi_{i,j}^{-1}.$$

In other words,

$$\Delta_j \circ h_j^{-1} = \left(\pi_{A^{\Phi_{i,j}}}(\Delta_i \circ h_i^{-1})\right) \circ \Phi_{i,j}^{-1}$$

and, therefore, the compatibility condition (4.4.1) holds. □

Now we turn to the proof of Theorem 1.4.6.

4.4 Proof of Theorem 1.4.6

Lemma 4.4.2 *Let $S \in \Pi$ be compactly supported. Let $g : \mathbb{H}^d \to \mathrm{GL}^+(2d, \mathbb{R})$ be smooth and constant outside some ball. Let ν be a smooth positive density on $\mathbb{H}^d$. We have*

$$S(1+\Delta_{g,\nu})^{-d-1} \in \mathcal{L}_{1,\infty},$$

$$\varphi(S(1+\Delta_{g,\nu})^{-d-1}) = c_d \Big(\int_{\mathbb{H}^d} \otimes \tau\Big)\Big(\mathrm{sym}(S) \cdot e^{\sum_{k_1,k_2=1}^{2d}(g^{-1})_{k_1 k_2} \otimes X_{k_1} X_{k_2}}\Big).$$

The operators are understood on $L_2(\mathbb{H}^d, \nu)$.

Proof Let f be the Radon-Nikodym derivative of ν. The operator $U = M_{f^{-\frac12}} : L_2(\mathbb{H}^d) \to L_2(\mathbb{H}^d, \nu)$ is unitary. Let $P = M_{f^{\frac12}} \Delta_{g,\nu} M_{f^{-\frac12}}$ be a differential operator on $L_2(\mathbb{H}^d)$. Since P is elliptic, formally self-adjoint and formally positive, it follows from Theorem 2.5.2 that $P : W^{2,2}(\mathbb{H}^d) \to L_2(\mathbb{H}^d)$ is self-adjoint and positive.

We have

$$U^{-1} S(1+\Delta_{g,\nu})^{-d-1} U = M_{f^{\frac12}} S M_{f^{-\frac12}} (1+P)^{-d-1}.$$

Since the operator S is compactly supported, it follows that $M_{f^{\frac12}} S M_{f^{-\frac12}} \in \Pi$ is also compactly supported. By Theorem 2.6.1, we have

$$M_{f^{\frac12}} S M_{f^{-\frac12}} (1+P)^{-d-1} \in \mathcal{L}_{1,\infty}.$$

Thus,

$$S(1+\Delta_{g,\nu})^{-d-1} \in \mathcal{L}_{1,\infty}.$$

We write

$$\varphi(S(1+\Delta_{g,\nu})^{-d-1}) = \varphi(M_{f^{\frac12}} S M_{f^{-\frac12}} (1+P)^{-d-1}),$$

where the operator on the left (respectively, on the right) hand side is understood on $L_2(\mathbb{H}^d, \nu)$ (respectively, on $L_2(\mathbb{H}^d)$).

Note that the principal part of P is $-\sum_{k_1,k_2=1}^{2d} X_{k_1} M_{(g^{-1})_{k_1,k_2}} X_{k_2}$. Using Theorem 2.6.1, we write

$$\varphi(M_{f^{\frac12}} S M_{f^{-\frac12}} (1+P)^{-d-1})$$
$$= c_d \Big(\int_{\mathbb{H}^d} \otimes \tau\Big)\Big(\mathrm{sym}(M_{f^{\frac12}} S M_{f^{-\frac12}}) \cdot e^{\sum_{k_1,k_2=1}^{2d}(g^{-1})_{k_1 k_2} \otimes X_{k_1} X_{k_2}}\Big).$$

Since S is compactly supported, it follows that

$$\operatorname{sym}(M_{f^{\frac12}} S M_{f^{-\frac12}}) = \operatorname{sym}(S).$$

This completes the proof. □

Lemma 4.4.3 *Let $(\mathcal{U}, h)$ be a chart and let $A : L_2(X) \to W^{2,2}(X)$ be a bounded linear mapping. If A is compactly supported in $\mathcal{U}$, then $A \in \mathcal{L}_{d+1,\infty}$.*

Proof Define the unitary operator $W : L_2(\mathcal{U}, \nu) \to L_2(h(\mathcal{U}), \nu \circ h^{-1})$ by setting $Wf = f \circ h$.

Since A is compactly supported in $\mathcal{U}$, it follows that the operator $WAW^{-1} : L_2(h(\mathcal{U})) \to W^{2,2}(h(\mathcal{U}))$ is bounded and compactly supported in $h(\mathcal{U})$. Set

$$B = \operatorname{Ext}_{h(\mathcal{U})}(WAW^{-1}).$$

We have that $B : L_2(\mathbb{H}^d) \to W^{2,2}(\mathbb{H}^d)$ is a bounded and compactly supported operator. Hence, $(1+\Delta)B : L_2(\mathbb{H}^d) \to L_2(\mathbb{H}^d)$ is a bounded operator.

Let $\psi \in C_c^\infty(\mathbb{H}^d)$ be such that $B = M_\psi B$. We have

$$B = M_\psi B = M_\psi (1+\Delta)^{-1} \cdot (1+\Delta)B.$$

The second factor on the right hand side is bounded by the preceding paragraph. The first factor on the right hand side belongs to $\mathcal{L}_{d+1,\infty}$ by Cwikel estimates. Hence, $B \in \mathcal{L}_{d+1,\infty}(L_2(\mathbb{H}^d))$. In other words,

$$WAW^{-1} \in \mathcal{L}_{d+1,\infty}(L_2(h(\mathcal{U}))).$$

Since WAW^{-1} is compactly supported in $h(\mathcal{U})$, the assertion follows. □

By Theorem 1.4.3, the operator $1 + \Delta_{G,\nu}$ is invertible in $L_2(X, \nu)$. Moreover, the inverse defines a bounded operator

$$(1+\Delta_{G,\nu})^{-1} : L_2(X, \nu) \to W^{2,2}(X). \tag{4.4.2}$$

Lemma 4.4.4 *Let $(\mathcal{U}, h)$ be a chart and let $\phi \in C^\infty(X)$ be compactly supported in $\mathcal{U}$. We have*

$$M_\phi (1+\Delta_{G,\nu})^{-1} \in \mathcal{L}_{d+1,\infty}.$$

Proof Choose $\psi \in C^\infty(X)$ compactly supported in $\mathcal{U}$ such that $\phi\psi = \phi$. We write

$$M_\phi (1+\Delta_{G,\nu})^{-1} = M_\phi (1+\Delta_{G,\nu})^{-1} M_\psi + M_\phi [M_\psi, (1+\Delta_{G,\nu})^{-1}]$$

$$= M_\phi (1+\Delta_{G,\nu})^{-1} M_\psi + M_\phi (1+\Delta_{G,\nu})^{-1}$$

$$[\Delta_{G,\nu}, M_\psi](1+\Delta_{G,\nu})^{-1}.$$

4.4 Proof of Theorem 1.4.6

Choose $\chi \in C^\infty(X)$ compactly supported in $\mathcal{U}$ such that $\psi \chi = \psi$. Since

$$[\Delta_{G,\nu}, M_\psi] = M_\chi [\Delta_{G,\nu}, M_\psi],$$

it follows that

$$M_\phi (1 + \Delta_{G,\nu})^{-1}$$
$$= M_\phi (1 + \Delta_{G,\nu})^{-1} M_\psi + M_\phi (1 + \Delta_{G,\nu})^{-1} M_\chi [\Delta_{G,\nu}, M_\psi](1 + \Delta_{G,\nu})^{-1}.$$

It is clear that

$$[\Delta_{G,\nu}, M_\psi] : W^{2,2}(X) \to L_2(X).$$

Hence, by (4.4.2), the operator

$$[\Delta_{G,\nu}, M_\psi](1 + \Delta_{G,\nu})^{-1}$$

is a well defined bounded operator on $L_2(X, \nu)$.

By (4.4.2), we also have that

$$M_\phi (1 + \Delta_{G,\nu})^{-1} M_\psi, M_\phi (1 + \Delta_{G,\nu})^{-1} M_\chi : L_2(X, \nu) \to W^{2,2}(X)$$

are bounded operators compactly supported in $\mathcal{U}$. By Lemma 4.4.3,

$$M_\phi (1 + \Delta_{G,\nu})^{-1} M_\psi, M_\phi (1 + \Delta_{G,\nu})^{-1} M_\chi \in \mathcal{L}_{d+1,\infty}.$$

The assertion follows now by combining the preceding paragraphs. $\square$

Proof of Theorem 1.4.6 (i) Let $(\phi_n)_{n=1}^N \subset C_c^\infty(X)$ be a good partition of unity. We write

$$(1 + \Delta_{G,\nu})^{-1} = \sum_{n=1}^N M_{\phi_n}(1 + \Delta_{G,\nu})^{-1}.$$

The assertion follows from Lemma 4.4.4. $\square$

Lemma 4.4.5 *For every $m \in \mathbb{N}$, $(1 + \Delta_{G,\nu})^{-m}$ is a bounded operator from $L_2(X)$ to $W^{2m,2}(X)$.*

Proof It is immediate that $(1 + \Delta_{G,\nu})^m : W^{2m,2}(X) \to L_2(X)$. Let us prove that the latter mapping is surjective. Indeed, otherwise there exists $u \in L_2(X)$ such that

$$\langle u, (1 + \Delta_{G,\nu})^m \psi \rangle = 0, \quad \psi \in C^\infty(X).$$

Fix a chart $(\mathcal{U}_i, h_i)$, denote $\Omega_i = h_i(\mathcal{U}_i)$ and define a distribution u_i on Ω_i by setting

$$\langle u_i, \phi \rangle = \langle u, \phi \circ h_i \rangle, \quad \phi \in C_c^\infty(\Omega_i).$$

By (4.2.2),

$$(1 + \Delta_{G,v})^m(\phi \circ h_i) = \left((1 + \Delta_{g_i, v \circ h_i^{-1}})^m \phi\right) \circ h_i.$$

Thus,

$$\begin{aligned}\langle (1 + \Delta_{g_i, v \circ h_i^{-1}})^m u_i, \phi \rangle &= \langle u_i, (1 + \Delta_{g_i, v \circ h_i^{-1}})^m \phi \rangle \\ &= \langle u, ((1 + \Delta_{g_i, v \circ h_i^{-1}})^m \phi) \circ h_i \rangle \\ &= \langle u, (1 + \Delta_{G,v})^m (\phi \circ h_i) \rangle \\ &= 0\end{aligned}$$

for every $\phi \in C_c^\infty(\Omega_i)$. In other words, u_i is a weak solution of the equation

$$(1 + \Delta_{g_i, v \circ h_i^{-1}})^m u_i = 0.$$

Applying Lemma 2.5.3 m times, we obtain that u_i is smooth.

Let $(\phi_n)_{n=1}^N$ be a good partition of unity. Let ϕ_n be compactly supported in $\mathcal{U}_{i_n}$. Since u_{i_n} is smooth, u is smooth in $\mathcal{U}_{i_n}$ (for every $1 \leq n \leq N$). Thus, $u \in C^\infty(X)$. Hence,

$$\langle (1 + \Delta_{G,v})^m u, \psi \rangle = \langle u, (1 + \Delta_{G,v})^m \psi \rangle = 0, \quad \psi \in C^\infty(X).$$

It follows that $(1 + \Delta_{G,v})^m u = 0$. Since $\ker((1 + \Delta_{G,v})^m) = 0$, the assertion follows. □

Lemma 4.4.6 *Let $\phi \in C^\infty(X)$. We have*

$$(1 + \Delta_{G,v})^{d+1} M_\phi (1 + \Delta_{G,v})^{-d-1} M_\phi - M_{\phi^2} \in \mathcal{L}_{2d+2, \infty}.$$

Proof We write

$$\begin{aligned}&(1 + \Delta_{G,v})^{d+1} M_\phi (1 + \Delta_{G,v})^{-d-1} M_\phi - M_{\phi^2} \\ &= [(1 + \Delta_{G,v})^{d+1}, M_\phi] \cdot (1 + \Delta_{G,v})^{-d-1} M_\phi \\ &= [(1 + \Delta_{G,v})^{d+1}, M_\phi](1 + \Delta_{G,v})^{-d-\frac{1}{2}} \cdot (1 + \Delta_{G,v})^{-\frac{1}{2}} M_\phi.\end{aligned}$$

4.4 Proof of Theorem 1.4.6

By Lemma 4.4.5, we have

$$(1+\Delta_{G,\nu})^{-d}: L_2(X) \to W^{2d,2}(X), \quad (1+\Delta_{G,\nu})^{-d-1}: L_2(X) \to W^{2d+2,2}(X).$$

By complex interpolation,

$$(1+\Delta_{G,\nu})^{-d-\frac12}: L_2(X) \to W^{2d+1,2}(X).$$

Note that $[(1+\Delta_{G,\nu})^{d+1}, M_\phi]$ is a differential operator of order $2d+1$ on X. Hence,

$$[(1+\Delta_{G,\nu})^{d+1}, M_\phi] \cdot (1+\Delta_{G,\nu})^{-d-\frac12}$$

is bounded. By Theorem 1.4.6 (i), $(1+\Delta_{G,\nu})^{-\frac12} \in \mathcal{L}_{2d+2,\infty}$ and the assertion follows. □

Lemma 4.4.7 *Suppose that a chart $(\mathcal{U}_i, h_i)$ satisfies the following conditions:*

(i) *Ω_i is a (Euclidean) ball in $\mathbb{H}^d$;*
(ii) *g_i is smooth on $\overline{\Omega_i}$;*
(iii) *g_i^{-1} is bounded from below (by a strictly positive scalar matrix) on Ω_i;*
(iv) *the Radon-Nikodym derivative of $\nu \circ h_i^{-1}$ is smooth on $\overline{\Omega_i}$;*
(v) *the Radon-Nikodym derivative of $\nu \circ h_i^{-1}$ is bounded from below by a strictly positive constant on Ω_i;*

Let $S \in \Pi_X$ be compactly supported in a chart $(\mathcal{U}_i, h_i)$. We have

$$\varphi(S(1+\Delta_{G,\nu})^{-d-1})$$
$$= c_d \Big(\int_{\Omega_i} \otimes \tau \Big) \Big(\mathrm{sym}(\mathrm{Ext}_{\Omega_i}(W_i S W_i^{-1})) \cdot e^{\sum_{k_1,k_2=1}^{2d}(g_i^{-1})_{k_1k_2} \otimes X_{k_1} X_{k_2}} \Big).$$

Proof Extend g_i to a smooth mapping $g_i : \mathbb{H}^d \to \mathrm{GL}^+(2d, \mathbb{R})$ and assume that g_i is constant outside some ball. Extend $\nu \circ h_i^{-1}$ to a smooth positive density on $\mathbb{H}^d$. Recall that $W_i : L_2(\mathcal{U}_i, \nu) \to L_2(\Omega_i, \nu \circ h_i)$ is the unitary operator given by the formula $W_i \xi = \xi \circ h_i^{-1}$. Extend W_i to an isometry $W_i : L_2(\mathcal{U}_i, \nu) \to L_2(\mathbb{H}^d, \nu_i)$ by the same formula $W_i \xi = \xi \circ h_i^{-1}$. We use this extended version of W_i everywhere in this proof, except for the very last display.

Choose $\phi \in C_c^\infty(\mathcal{U}_i)$ such that $S = S M_\phi = M_\phi S$. By the tracial property, we have

$$\varphi(S(1+\Delta_{G,\nu})^{-d-1}) = \varphi(M_\phi S M_\phi (1+\Delta_{G,\nu})^{-d-1}) = \varphi(S M_\phi (1+\Delta_{G,\nu})^{-d-1} M_\phi).$$

Applying the isometry W_i, we obtain

$$\varphi(S(1+\Delta_{G,\nu})^{-d-1}) = \varphi(W_i S M_\phi (1+\Delta_{G,\nu})^{-d-1} M_\phi W_i^*),$$

where the operator on the right hand side is considered as an operator in $L_2(\mathbb{H}^d, \nu_i)$.

Choose $\psi \in C_c^\infty(\mathbb{H}^d)$ compactly supported in Ω_i and such that $\psi \cdot (\phi \circ h_i^{-1}) = \phi \circ h_i^{-1}$. We now write

$$W_i S M_\phi (1 + \Delta_{G,v})^{-d-1} M_\phi W_i^*$$
$$= W_i S W_i^* \cdot M_\psi \cdot W_i M_\phi (1 + \Delta_{G,v})^{-d-1} M_\phi W_i^*$$
$$= W_i S W_i^* (1 + \Delta_{g_i,v_i})^{-d-1} \cdot (1 + \Delta_{g_i,v_i})^{d+1} M_\psi \cdot W_i M_\phi (1 + \Delta_{G,v})^{-d-1} M_\phi W_i^*.$$

Since $(1 + \Delta_{g_i,v_i})^{d+1} M_\psi$ is a differential operator compactly supported in Ω_i, it follows that

$$(1+\Delta_{g_i,v_i})^{d+1} M_\psi W_i \cdot M_\phi = W_i(1+\Delta_{G,v})^{d+1} M_{\psi \circ h_i} \cdot M_\phi = W_i(1+\Delta_{G,v})^{d+1} M_\phi.$$

Thus,

$$W_i S M_\phi (1 + \Delta_{G,v})^{-d-1} M_\phi W_i^*$$
$$= W_i S W_i^* (1 + \Delta_{g_i,v_i})^{-d-1} \cdot W_i (1 + \Delta_{G,v})^{d+1} M_\phi (1 + \Delta_{G,v})^{-d-1} M_\phi W_i^*.$$

The operator $W_i S W_i^* \in \Pi$ is compactly supported. By Lemma 4.4.2, we have

$$W_i S W_i^* (1 + \Delta_{g_i,v_i})^{-d-1} \in \mathcal{L}_{1,\infty}.$$

By Lemma 4.4.6, we have

$$W_i (1 + \Delta_{G,v})^{d+1} M_\phi (1 + \Delta_{G,v})^{-d-1} M_\phi W_i^* - W_i M_{\phi^2} W_i^* \in \mathcal{L}_{2d+2,\infty}.$$

Thus,

$$W_i S M_\phi (1 + \Delta_{G,v})^{-d-1} M_\phi W_i^* - W_i S W_i^* (1 + \Delta_{g_i,v_i})^{-d-1} \cdot W_i M_{\phi^2} W_i^* \in \mathcal{L}_{\frac{2d+2}{2d+3},\infty}.$$

Since φ vanishes on $\mathcal{L}_{\frac{2d+2}{2d+3},\infty}$, it follows that

$$\varphi(S(1 + \Delta_{G,v})^{-d-1}) = \varphi(W_i S W_i^* (1 + \Delta_{g_i,v_i})^{-d-1} \cdot W_i M_{\phi^2} W_i^*)$$
$$= \varphi(W_i M_{\phi^2} W_i^* \cdot W_i S W_i^* (1 + \Delta_{g_i,v_i})^{-d-1})$$
$$= \varphi(W_i M_{\phi^2} S W_i^* (1 + \Delta_{g_i,v_i})^{-d-1})$$
$$= \varphi(W_i S W_i^* (1 + \Delta_{g_i,v_i})^{-d-1}).$$

4.4 Proof of Theorem 1.4.6

It follows now from Lemma 4.4.2 that

$$\varphi(S(1+\Delta_{G,\nu})^{-d-1})$$
$$= c_d \Big(\int_{\mathbb{H}^d} \otimes \tau \Big) \Big(\text{sym}(W_i S W_i^*) \cdot e^{\sum_{k_1,k_2=1}^{2d} (g_i^{-1})_{k_1 k_2} \otimes X_{k_1} X_{k_2}} \Big).$$

Clearly,

$$W_i S W_i^* = \text{Ext}_{\Omega_i}(W_i S W_i^{-1}),$$

where, in the right hand side, W_i is understood as a unitary operator from $L_2(\mathcal{U}_i, \nu)$ to $L_2(\Omega_i, \nu \circ h_i)$. $\square$

Lemma 4.4.8 *Let $S \in \Pi_X$ be compactly supported in a chart $(\mathcal{U}_i, h_i)$. We have*

$$\varphi(S(1+\Delta_{G,\nu})^{-d-1})$$
$$= c_d \Big(\int_{\Omega_i} \otimes \tau \Big) \Big(\text{sym}(\text{Ext}_{\Omega_i}(W_i S W_i^{-1})) \cdot e^{\sum_{k_1,k_2=1}^{2d} (g_i^{-1})_{k_1 k_2} \otimes X_{k_1} X_{k_2}} \Big).$$

Proof Let S be supported in a compact set $K \subset \mathcal{U}_i$. For every $x \in h_i(K)$, set $r(x) = \frac{1}{2}\text{dist}(x, \partial \Omega_i)$ and consider the Euclidean ball $B(x, r(x))$. By compactness, there is a finite collection $\{x_l\}_{l=1}^{L} \subset h_i(K)$ such that $h_i(K) \subset \cup_{l=1}^{L} B(x_l, r(x_l))$. For every $1 \leq l \leq L$, choose $0 \leq \phi_l \in C_c(B(x_l, r(x_l)))$ such that $\sum_{l=1}^{L} \phi_l = 1$ on $h_i(K)$. We have

$$S = \Big(\sum_{l=1}^{L} M_{\phi_l \circ h_i} \Big) S = \sum_{l=0}^{L} S_l,$$

where

$$S_l = M_{(\phi_l \circ h_i)^{\frac{1}{2}}} S M_{(\phi_l \circ h_i)^{\frac{1}{2}}}, \quad 1 \leq l \leq L, \quad S_0 = \sum_{m=1}^{L} M_{(\phi_m \circ h_i)^{\frac{1}{2}}} [M_{(\phi_m \circ h_i)^{\frac{1}{2}}}, S].$$

For $1 \leq l \leq L$, consider the chart $\{h_i^{-1}(B(x_l, r(x_l))), h_i\}$. This chart satisfies the conditions imposed in Lemma 4.4.7. Clearly, S_l is compactly supported in this chart. Hence,

$$\varphi(S_l(1+\Delta_{G,\nu})^{-d-1})$$
$$= c_d \Big(\int_{B(x_l, r(x_l))} \otimes \tau \Big) \Big(\text{sym}(\text{Ext}_{B(x_l, r(x_l))}(W_i S_l W_i^{-1})) \cdot e^{\sum_{k_1,k_2=1}^{2d} (g_i^{-1})_{k_1 k_2} \otimes X_{k_1} X_{k_2}} \Big)$$
$$= c_d \Big(\int_{\Omega_i} \otimes \tau \Big) \Big(\text{sym}(\text{Ext}_{\Omega_i}(W_i S_l W_i^{-1})) \cdot e^{\sum_{k_1,k_2=1}^{2d} (g_i^{-1})_{k_1 k_2} \otimes X_{k_1} X_{k_2}} \Big).$$

Now, consider the case $l = 0$. The operator S_0 is compact. Hence,

$$S_0(1 + \Delta_{G,v})^{-d-1} \in (\mathcal{L}_{1,\infty})_0$$

and

$$\varphi(S_0(1 + \Delta_{G,v})^{-d-1}) = 0.$$

We now write

$$\varphi(S(1 + \Delta_{G,v})^{-d-1})$$

$$= \sum_{l=0}^{L} \varphi(S_l(1 + \Delta_{G,v})^{-d-1})$$

$$= c_d \sum_{l=1}^{L} (\int_{\Omega_i} \otimes \tau) \left(\text{sym}(\text{Ext}_{\Omega_i}(W_i S_l W_i^{-1})) \cdot e^{\sum_{k_1,k_2=1}^{2d}(g_i^{-1})_{k_1 k_2} \otimes X_{k_1} X_{k_2}} \right)$$

$$= c_d (\int_{\Omega_i} \otimes \tau) \left(\text{sym}(\text{Ext}_{\Omega_i}(W_i (\sum_{l=1}^{L} S_l) W_i^{-1})) \cdot e^{\sum_{k_1,k_2=1}^{2d}(g_i^{-1})_{k_1 k_2} \otimes X_{k_1} X_{k_2}} \right).$$

Clearly, $\sum_{l=1}^{L} S_l = S - S_0$. Since S_0 is compact, it follows that

$$\text{sym}(\text{Ext}_{\Omega_i}(W_i (\sum_{l=1}^{L} S_l) W_i^{-1})) = \text{sym}(\text{Ext}_{\Omega_i}(W_i S W_i^{-1})).$$

This completes the proof. □

Next, we rewrite the right hand side of the formula in Lemma 4.4.8 in terms of globally defined data.

Lemma 4.4.9 *Let $S \in \Pi_X$ be compactly supported in a chart $(\mathcal{U}_i, h_i)$. We have*

$$\varphi(S(1 + \Delta_{G,v})^{-d-1}) = c_d \Lambda(\text{sym}_X(S) \cdot e^{-q_X}).$$

Proof The assertion follows from the definition of the Liouville trace Λ on $L_\infty(E)$ given in Theorem 1.4.4, the definition of q_X given in Lemmas 1.4.5 and 4.4.8. □

It remains to glue these local formulas together into a global one, using a partition of unity.

Proof of Theorem 1.4.6 (ii) Let $S \in \Pi_X$ and let $(\phi_n)_{n=1}^N$ be a fixed good partition of unity. We write

$$S = \sum_{n=0}^N S_n, \quad S_0 = \sum_{m=1}^N M_{\phi_m}^{\frac{1}{2}} \cdot [M_{\phi_m}^{\frac{1}{2}}, S], \quad S_n = M_{\phi_n}^{\frac{1}{2}} S M_{\phi_n}^{\frac{1}{2}}, \quad n \geq 1.$$

By assumption, $[S, M_\psi]$ is compact for every $\psi \in C(X)$. In particular, S_0 is compact. Thus,

$$\varphi(S_0(1 + \Delta_G)^{-d-1}) = 0, \quad \mathrm{sym}_X(S_0) = 0.$$

By Lemma 4.4.9, we have

$$\varphi(S(1 + \Delta_G)^{-d-1}) = \sum_{n=0}^N \varphi(S_n(1 + \Delta_G)^{-d-1})$$

$$= \sum_{n=1}^N c_d \Lambda(\mathrm{sym}_X(S_n) e^{-q_X})$$

$$= c_d \Lambda(\mathrm{sym}_X(S) e^{-q_X}).$$

This completes the proof. □

4.5 Spectrally Correct Sub-Riemannian Volume

In this section, we prove the concrete expression for the sub-Riemannian volume vol_G given by the formula (1.4.4).

As before, τ denotes the natural trace on $\mathrm{VN}(\mathbb{H}^d)$ (see Sect. 2.1.3).

Lemma 4.5.1 *Let $D \in \mathrm{GL}^+(2d, \mathbb{R})$ be diagonal with $d_{i,i} = d_{i+d,i+d}$ for $1 \leq i \leq d$. We have*

$$\tau\Big(\exp(\sum_{l=1}^{2d} d_{l,l} X_l^2)\Big) = 2^{1-d} \int_0^\infty \prod_{l=1}^d \frac{s}{\sinh(d_{l,l}s)} ds.$$

Proof By Proposition 2.1.4, we have

$$\pi\Big(-\sum_{l=1}^d d_{l,l}(p_l^2 + q_l^2) \otimes |s|\Big) = \sum_{l=1}^{2d} d_{l,l} X_l^2$$

and

$$\tau\Big(\exp(\sum_{l=1}^{2d} d_{l,l} X_l^2)\Big) = \int_{\mathbb{R}} \text{Tr}_{L_2(\mathbb{R}^d)}(\exp(-\sum_{l=1}^{d} d_{l,l}|s|(p_l^2+q_l^2))) \cdot |s|^d ds.$$

Now we proceed as follows:

$$\text{Tr}_{L_2(\mathbb{R}^d)}(\exp(-\sum_{l=1}^{d} d_{l,l}|s|(p_l^2+q_l^2))) = \prod_{l=1}^{d} \text{Tr}_{L_2(\mathbb{R})}(\exp(-d_{l,l}|s|(p^2+q^2)))$$

$$= \prod_{l=1}^{d} \sum_{n \geq 0} \exp(-d_{l,l}|s|(2n+1))$$

$$= \prod_{l=1}^{d} \frac{e^{-d_{l,l}|s|}}{1-e^{-2d_{l,l}|s|}}$$

$$= 2^{-d} \prod_{l=1}^{d} \frac{1}{\sinh(d_{l,l}|s|)}.$$

The assertion follows immediately. □

We now employ Williamson theorem—see Theorem 2.5.4.

Lemma 4.5.2 *If $A \in \text{GL}^+(2d, \mathbb{R})$, then*

$$\tau\Big(\exp(\sum_{k_1,k_2=1}^{2d} a_{k_1 k_2} X_{k_1} X_{k_2})\Big) = 2^{1-d} \det(A)^{-\frac{1}{2}} \cdot \beth(A).$$

Proof Let S and D be as in Theorem 2.5.4. We have

$$a_{k_1 k_2} = \sum_{l=1}^{2d} s_{l,k_1} d_{l,l} s_{l,k_2}.$$

Thus,

$$\sum_{k_1,k_2=1}^{2d} a_{k_1 k_2} X_{k_1} X_{k_2} = \sum_{k_1,k_2=1}^{2d} \sum_{l=1}^{2d} s_{l,k_1} d_{l,l} s_{l,k_2} X_{k_1} X_{k_2} = \sum_{l=1}^{2d} d_{l,l} \Big(\sum_{k=1}^{2d} s_{l,k} X_k\Big)^2.$$

Set $B = S \oplus 1_{\mathbb{C}} \in \text{Aut}(\mathbb{H}^d)$. By Lemma 2.1.7, we have

$$V_B^{-1} X_l V_B = \sum_{k=1}^{2d} s_{l,k} X_k, \quad 1 \leq l \leq d.$$

4.5 Spectrally Correct Sub-Riemannian Volume

Hence,

$$\exp(\sum_{k_1,k_2=1}^{2d} a_{k_1 k_2} X_{k_1} X_{k_2}) = \pi_B\left(\exp(\sum_{l=1}^{2d} d_{l,l} X_l^2)\right).$$

Since $\det(S) = 1$, it follows from Lemma 4.3.1 that π_B preserves τ. Hence,

$$\tau\left(\exp(\sum_{k_1,k_2=1}^{2d} a_{k_1 k_2} X_{k_1} X_{k_2})\right) = \tau\left(\exp(\sum_{l=1}^{2d} d_{l,l} X_l^2)\right).$$

By Lemma 4.5.1, we get

$$\tau\left(\exp(\sum_{k_1,k_2=1}^{2d} a_{k_1 k_2} X_{k_1} X_{k_2})\right) = 2^{1-d} \prod_{l=1}^{d} d_{l,l}^{-1} \cdot \int_0^\infty \prod_{l=1}^{d} \frac{d_{l,l} s}{\sinh(d_{l,l} s)} ds.$$

Now, let $C \in \operatorname{GL}^+(2d, \mathbb{R})$ be the diagonal matrix with $c_{i,i} = d_{i,i}$ for $1 \leq i \leq d$ and $c_{i,i} = -d_{i,i}$ for $d+1 \leq i \leq 2d$. We have

$$\prod_{l=1}^{d} \frac{d_{l,l} s}{\sinh(d_{l,l} s)} = \det^{\frac{1}{2}}\left(\frac{sC}{\sinh(sC)}\right), \quad \prod_{l=1}^{d} d_{l,l}^{-1} = |\det(C)|^{-\frac{1}{2}}.$$

By Theorem 2.5.4, C is similar to $\iota\Omega A$. Hence,

$$\prod_{l=1}^{d} \frac{d_{l,l} s}{\sinh(d_{l,l} s)} = \det^{\frac{1}{2}}\left(\frac{\iota s \Omega A}{\sinh(\iota s \Omega A)}\right), \quad \prod_{l=1}^{d} d_{l,l}^{-1} = |\det(C)|^{-\frac{1}{2}} = \det(A)^{-\frac{1}{2}}.$$

Thus,

$$\tau\left(\exp(\sum_{k_1,k_2=1}^{2d} a_{k_1 k_2} X_{k_1} X_{k_2})\right) = 2^{1-d} \det(A)^{-\frac{1}{2}} \cdot \int_0^\infty \det^{\frac{1}{2}}\left(\frac{\iota s \Omega A}{\sinh(\iota s \Omega A)}\right) ds.$$

The assertion follows now from the definition of $\beth$. □

Proof of Theorem 1.4.8 By (1.4.3), (1.4.1) and Lemma 1.4.5, for any $f \in C_c(\mathcal{U}_i)$, we have

$$\int_X f d\operatorname{vol}_G = \Lambda(f e^{-qX})$$

$$= \int_{\Omega_i} \tau\left(f(h_i^{-1}(p)) e^{\sum_{k_1,k_2=1}^{2d} (g_i^{-1})_{k_1,k_2}(p) X_{k_1} X_{k_2}}\right) dp$$

$$= \int_{\Omega_i} f(h_i^{-1}(p)) \cdot \tau\left(e^{\sum_{k_1,k_2=1}^{2d} (g_i^{-1})_{k_1,k_2}(p) X_{k_1} X_{k_2}}\right) dp.$$

The assertion follows now from Lemma 4.5.2. □

References

1. Alberti, P., Matthes, R.: Connes' trace formula and Dirac realization of Maxwell and Yang-Mills action. In: Noncommutative Geometry and the Standard Model of Elementary Particle Physics (Hesselberg, 1999). Lecture Notes in Physics, vol. 596, pp. 40–74. Springer, Berlin (2002)
2. Arnold, V.: Mathematical Methods of Classical Mechanics. Graduate Texts in Mathematics, vol. 60. Springer, New York (1978)
3. Atiyah, M., Singer, I.: The index of elliptic operators. I. Ann. Math. **87**, 484–530 (1968)
4. Barilari, D., Rizzi, L.: A formula for Popp's volume in sub-Riemannian geometry. Anal. Geom. Metr. Spaces **1**, 42–57 (2013)
5. Baum, P., Douglas, R.: Toeplitz operators and Poincare duality. Toeplitz centennial (Tel Aviv, 1981). In: Operator Theory: Advances and Applications, vol. 4, pp. 137–166. Birkhäuser, Basel (1982)
6. Beals, R.: Characterization of pseudodifferential operators and applications. Duke Math. J. **44**(1), 45–57 (1977)
7. Beals, R., Greiner, P.: Calculus on Heisenberg Manifolds. Annals of Mathematics Studies, vol. 119. Princeton University Press, Princeton (1988)
8. Beals, R., Gaveau, B., Greiner, P.: Hamilton-Jacobi theory and the heat kernel on Heisenberg groups. J. Math. Pures Appl. **79**(7), 633–689 (2000)
9. Birman, M., Solomyak, M.: Asymptotic behavior of the spectrum of pseudodifferential operators with anisotropically homogeneous symbols. Vestnik Leningrad. Univ. **13**, 13–21, 169 (1977)
10. Blackadar, B.: Operator algebras. In: Theory of C^*-Algebras and von Neumann Algebras. Encyclopaedia of Mathematical Sciences. Operator Algebras and Non-commutative Geometry, III, vol. 122. Springer, Berlin (2006)
11. Christ, M., Geller, D., Glowacki, P., Polin, L. Pseudodifferential operators on groups with dilations. Duke Math. J. **68**(1), 31–65 (1992)
12. Colin de Verdiére, Y., Hillairet, L., Trélat, E.: Spectral asymptotics for sub-Riemannian Laplacians, I: quantum ergodicity and quantum limits in the 3-dimensional contact case. Duke Math. J. **167**(1), 109–174 (2018)
13. Colin de Verdiére, Y., Hillairet, L., Trélat, E.: Spectral asymptotics for sub-Riemannian Laplacians (2022). arXiv:2212.02920
14. Connes, A.: The action functional in noncommutative geometry. Commun. Math. Phys. **117**(4), 673–683 (1988)

15. Cordes, H.: Spectral Theory of Linear Differential Operators and Comparison Algebras. London Mathematical Society Lecture Note Series, vol. 76. Cambridge University Press, Cambridge (1987)
16. Couchet, N., Yuncken, R.: On polyhomogeneous symbols and the Heisenberg pseudodifferential calculus (2022). arXiv:2210.15391
17. Couchet, N., Yuncken, R.: A groupoid approach to the Wodzicki residue. J. Funct. Anal. **286**(4), 110268, 24pp. (2024)
18. Dave, S., Haller, S.: The heat asymptotics on filtered manifolds. J. Geom. Anal. **30**(1), 337–389 (2020)
19. Debord, C., Skandalis, G.: Adiabatic groupoid, crossed product by $\mathbb{R}_+^*$ and pseudodifferential calculus. Adv. Math. **257**, 66–91 (2014)
20. de Gosson, M.: Symplectic Methods in Harmonic Analysis and in Mathematical Physics. Pseudo-Differential Operators. Theory and Applications, vol. 7. Birkhäuser, Basel (2011)
21. Dixmier, J.: C^*-Algebras. North-Holland Mathematical Library, vol. 15. North-Holland, Amsterdam (1977)
22. Dykema, K., Figiel, T., Weiss, G., Wodzicki, M.: Commutator structure of operator ideals. Adv. Math. **185**(1), 1–79 (2004)
23. Dynin, A.: Pseudodifferential operators on the Heisenberg group. Dokl. Akad. Nauk SSSR **225**, 1245–1248 (1975)
24. Dynin, A.: An algebra of pseudodifferential operators on the Heisenberg groups. Symbolic calculus. Dokl. Akad. Nauk SSSR **227**, 792–795 (1976)
25. Eliashberg, Y., Thurston, W.: Confoliations. University Lecture Series, vol. 13. American Mathematical Society, Providence (1998)
26. Epstein, C.: Lectures on indices and relative indices on contact and CR-manifolds. In: Woods Hole Mathematics, Series Knots Everything, vol. 34, pp. 27–93, World Science Publication, Hackensack (2004)
27. Epstein, C., Melrose R.: The Heisenberg algebra, index theory and homology. Unpublished book. https://math.mit.edu/~rbm/book.html
28. Fack, T., Kosaki, H.: Generalized s-numbers of τ-measurable operators. Pac. J. Math. **123**(2), 269–300 (1986)
29. Fan, Z., Li, J., McDonald, E., Sukochev, F., Zanin, D.: Endpoint weak Schatten class estimates and trace formula for commutators of Riesz transforms with multipliers on Heisenberg groups. J. Funct. Anal. **286**(1), 110188, 72pp. (2024)
30. Fermanian-Kammerer, C., Fischer, V.: Defect measures on graded Lie groups. Ann. Sc. Norm. Super. Pisa Cl. Sci. **21**, 207–291 (2020)
31. Fischer, V., Ruzhansky, M.: Quantization on Nilpotent Lie Groups. Progress in Mathematics, vol. 314. Birkhäuser, Basel (2016)
32. Folland, G.: Subelliptic estimates and function spaces on nilpotent Lie groups. Ark. Mat. **13**(2), 161–207 (1975)
33. Folland, G.: Harmonic Analysis in Phase Space. Annals of Mathematics Studies, vol. 122. Princeton University Press, Princeton (1989)
34. Folland, G., Stein, E.: Estimates for the $\bar{\partial}_b$ complex and analysis on the Heisenberg group. Commun. Pure Appl. Math. **27**, 429–522 (1974)
35. Garofalo, N., Nhieu, D.: Lipschitz continuity, global smooth approximations and extension theorems for Sobolev functions in Carnot-Caratheodory spaces. J. Anal. Math. **74**, 67–97 (1998)
36. Gracia-Bondia, J., Varilly, J., Figueroa, H.: Elements of Noncommutative Geometry. Birkhäuser Advanced Texts: Basler Lehrbücher. Birkhäuser, Boston (2001)
37. Helffer, B.: Théorie spectrale pour des opérateurs globalement elliptiques. In: Astérisque, vol. 112. Société Mathématique de France, Paris (1984)
38. Hörmander, L.: Hypoelliptic second order differential equations. Acta Math. **119**, 147–171 (1967)
39. Joshi, M.: Lectures on Pseudo-Differential Operators (1999). arXiv:math/9906155

References

40. Kadison, R., Ringrose, J.: Fundamentals of the Theory of Operator Algebras. Vol. I. Elementary Theory. Pure and Applied Mathematics, vol. 100. Academic Press, New York (1983)
41. Kohn, J., Nirenberg, L.: An algebra of pseudo-differential operators. Commun. Pure Appl. Math. **18**, 269–305 (1965)
42. Koranyi, A., Reimann, H.: Foundations for the theory of quasiconformal mappings on the Heisenberg group. Adv. Math. **111**(1), 1–87 (1995)
43. Kordyukov, Y., Sukochev, F., Zanin, D.: A C^*-algebraic approach to the principal symbol. III. J. Noncommut. Geom. **18**, 1265–1314 (2024)
44. Lord, S., Sukochev, F., Zanin, D.: Advances in Dixmier Traces and Applications. Advances in Noncommutative Geometry, pp. 491–583. Springer, Cham (2019)
45. Lord, S., Sukochev, F., Zanin, D.: Singular Traces. Vol. 1. Theory. De Gruyter Studies in Mathematics, vol. 46/1. De Gruyter, Berlin (2021)
46. Lord, S., Sukochev, F., Zanin, D., McDonald, E.: Singular Traces. Vol. 2. Trace Formulas. De Gruyter Studies in Mathematics, 46/2. De Gruyter, Berlin (2023)
47. McDonald, E., Sukochev, F., Zanin, D.: A C^*-algebraic approach to the principal symbol II. Math. Ann. **374**(1–2), 273–322 (2019)
48. McDonald, E., Sukochev, F., Zanin, D.: Spectral estimates and asymptotics for stratified Lie groups. J. Funct. Anal. **285**(10), 110105, 64pp. (2023)
49. Melo, S.: Norm closure of classical pseudodifferential operators does not contain Hörmander's class. In: Geometric Analysis of PDE and Several Complex Variables. Contemporary Mathematics, vol. 368, pp. 329–336. The American Mathematical Society, Providence (2005)
50. Melrose, R.: The Atiyah-Patodi-Singer Index Theorem. Research Notes in Mathematics, vol. 4. A K Peters, Wellesley (1993)
51. Michor, P.: Topics in Differential Geometry. American Mathematical Society, Providence (2008)
52. Pedersen, G.: C^*-Algebras and Their Automorphism Groups. Pure and Applied Mathematics (Amsterdam). Academic Press, London (2018)
53. Ponge, R.: Noncommutative residue for Heisenberg manifolds. Applications in CR and contact geometry. J. Funct. Anal. **252**(2), 399–463 (2007)
54. Ponge, R.: Heisenberg calculus and spectral theory of hypoelliptic operators on Heisenberg manifolds. Mem. Am. Math. Soc. **194**, 906 (2008)
55. Potapov, D., Sukochev, F.: Unbounded Fredholm modules and double operator integrals. J. Reine Angew. Math. **626**, 159–185 (2009)
56. Reed, M., Simon, B.: Methods of Modern Mathematical Physics. IV. Analysis of Operators. Academic Press, New York (1978)
57. Reed, M., Simon, B.: Methods of Modern Mathematical Physics. I. Functional Analysis, 2nd edn. Academic Press, New York (1980)
58. Sakai, S.: C^*-algebras and W^*-algebras. Reprint of the 1971 edition. Classics in Mathematics. Springer, Berlin (1998)
59. Sukochev, F., Zanin, D.: A C^*-algebraic approach to the principal symbol. I. J. Oper. Theory **80**(2), 481–522 (2018)
60. Takesaki, M.: Theory of Operator Algebras. II. Encyclopaedia of Mathematical Sciences, vol. 125. Operator Algebras and Non-commutative Geometry, 6. Springer, Berlin (2003)
61. Taylor, M.: Noncommutative microlocal analysis. I. Mem. Am. Math. Soc. **52**(313), iv+182 pp. (1984)
62. Taylor, M.: Microlocal Weyl formula on contact manifolds. Commun. Partial Differ. Equ. **45**(5), 392–413 (2020)
63. Treves, F.: Introduction to Pseudodifferential and Fourier Integral Operators. Vol. 1. Pseudodifferential Operators. University Series in Mathematics. Plenum Press, New York (1980)
64. van Erp, E.: The Atiyah-Singer formula for subelliptic operators on a contact manifold, Part I. Ann. Math. **171**, 1647–1681 (2010)

65. van Erp, E.: The Atiyah-Singer formula for subelliptic operators on a contact manifold, Part II. Ann. Math. **171**, 1683–1706 (2010)
66. van Erp, E., Yuncken, R.: A groupoid approach to pseudodifferential calculi. J. Reine Angew. Math. **756**, 151–182 (2019)
67. Zanin, D., Sukochev, F.: Connes integration formula: a constructive approach. Funct. Anal. Appl. **57**(1), 40–59 (2023)

LECTURE NOTES IN MATHEMATICS 🙟 Springer

Editors in Chief: J.-M. Morel, B. Teissier;

Editorial Policy

1. Lecture Notes aim to report new developments in all areas of mathematics and their applications – quickly, informally and at a high level. Mathematical texts analysing new developments in modelling and numerical simulation are welcome.

   Manuscripts should be reasonably self-contained and rounded off. Thus they may, and often will, present not only results of the author but also related work by other people. They may be based on specialised lecture courses. Furthermore, the manuscripts should provide sufficient motivation, examples and applications. This clearly distinguishes Lecture Notes from journal articles or technical reports which normally are very concise. Articles intended for a journal but too long to be accepted by most journals, usually do not have this "lecture notes" character. For similar reasons it is unusual for doctoral theses to be accepted for the Lecture Notes series, though habilitation theses may be appropriate.

2. Besides monographs, multi-author manuscripts resulting from SUMMER SCHOOLS or similar INTENSIVE COURSES are welcome, provided their objective was held to present an active mathematical topic to an audience at the beginning or intermediate graduate level (a list of participants should be provided).

   The resulting manuscript should not be just a collection of course notes, but should require advance planning and coordination among the main lecturers. The subject matter should dictate the structure of the book. This structure should be motivated and explained in a scientific introduction, and the notation, references, index and formulation of results should be, if possible, unified by the editors. Each contribution should have an abstract and an introduction referring to the other contributions. In other words, more preparatory work must go into a multi-authored volume than simply assembling a disparate collection of papers, communicated at the event.

3. Manuscripts should be submitted either online at www.editorialmanager.com/lnm to Springer's mathematics editorial in Heidelberg, or electronically to one of the series editors. Authors should be aware that incomplete or insufficiently close-to-final manuscripts almost always result in longer refereeing times and nevertheless unclear referees' recommendations, making further refereeing of a final draft necessary. The strict minimum amount of material that will be considered should include a detailed outline describing the planned contents of each chapter, a bibliography and several sample chapters. Parallel submission of a manuscript to another publisher while under consideration for LNM is not acceptable and can lead to rejection.

4. In general, **monographs** will be sent out to at least 2 external referees for evaluation.

   A final decision to publish can be made only on the basis of the complete manuscript, however a refereeing process leading to a preliminary decision can be based on a pre-final or incomplete manuscript.

   Volume Editors of **multi-author works** are expected to arrange for the refereeing, to the usual scientific standards, of the individual contributions. If the resulting reports can be forwarded to the LNM Editorial Board, this is very helpful. If no reports are forwarded or if other questions remain unclear in respect of homogeneity etc, the series editors may wish to consult external referees for an overall evaluation of the volume.

5. Manuscripts should in general be submitted in English. Final manuscripts should contain at least 100 pages of mathematical text and should always include

   - a table of contents;
   - an informative introduction, with adequate motivation and perhaps some historical remarks: it should be accessible to a reader not intimately familiar with the topic treated;
   - a subject index: as a rule this is genuinely helpful for the reader.
   - For evaluation purposes, manuscripts should be submitted as pdf files.

6. Careful preparation of the manuscripts will help keep production time short besides ensuring satisfactory appearance of the finished book in print and online. After acceptance of the manuscript authors will be asked to prepare the final LaTeX source files (see LaTeX templates online: https://www.springer.com/gb/authors-editors/book-authors-editors/manuscriptpreparation/5636) plus the corresponding pdf- or zipped ps-file. The LaTeX source files are essential for producing the full-text online version of the book, see http://link.springer.com/bookseries/304 for the existing online volumes of LNM). The technical production of a Lecture Notes volume takes approximately 12 weeks. Additional instructions, if necessary, are available on request from lnm@springer.com.

7. Authors receive a total of 30 free copies of their volume and free access to their book on SpringerLink, but no royalties. They are entitled to a discount of 33.3 % on the price of Springer books purchased for their personal use, if ordering directly from Springer.

8. Commitment to publish is made by a *Publishing Agreement*; contributing authors of multiauthor books are requested to sign a *Consent to Publish form*. Springer-Verlag registers the copyright for each volume. Authors are free to reuse material contained in their LNM volumes in later publications: a brief written (or e-mail) request for formal permission is sufficient.

Addresses:
Professor Jean-Michel Morel, CMLA, École Normale Supérieure de Cachan, France
E-mail: moreljeanmichel@gmail.com

Professor Bernard Teissier, Equipe Géométrie et Dynamique,
Institut de Mathématiques de Jussieu – Paris Rive Gauche, Paris, France
E-mail: bernard.teissier@imj-prg.fr

Springer: Ute McCrory, Mathematics, Heidelberg, Germany,
E-mail: lnm@springer.com

GPSR Compliance

The European Union's (EU) General Product Safety Regulation (GPSR) is a set of rules that requires consumer products to be safe and our obligations to ensure this.

If you have any concerns about our products, you can contact us on

ProductSafety@springernature.com

In case Publisher is established outside the EU, the EU authorized representative is:

Springer Nature Customer Service Center GmbH
Europaplatz 3
69115 Heidelberg, Germany

www.ingramcontent.com/pod-product-compliance
Ingram Content Group UK Ltd.
Pitfield, Milton Keynes, MK11 3LW, UK
UKHW021827210426